Natural and Living Biomaterials

Editors

Garth W. Hastings, D.Sc., Ph.D., C. Chem., F.R.S.C.
Head of Biomedical Engineering Unit
North Staffordshire Polytechnic, and
Honorary Scientific Officer
North Staffordshire Area Health District
Medical Institute
Hartshill, Stoke-on-Trent
England

Paul Ducheyne, Ph.D.
Associate Professor of Biomedical Engineering
University of Pennsylvania
Philadelphia, Pennsylvania

CRC Series in Structure-Property Relationships of Biomaterials

Series Editors-in-Chief

Garth W. Hastings, D.Sc., Ph.D., C. Chem., F.R.S.C.
Paul Ducheyne, Ph.D.

CRC Press
Taylor & Francis Group
Boca Raton London New York

CRC Press is an imprint of the
Taylor & Francis Group, an informa business

First published 1984 by CRC Press
Taylor & Francis Group
6000 Broken Sound Parkway NW, Suite 300
Boca Raton, FL 33487-2742

Reissued 2018 by CRC Press

Library of Congress Cataloging in Publication Data
Main entry under title:

Natural and living biomaterials.
Bibliography: p.

1. Bones. 2. Ligaments. 3. Tendons. 3. Biomedical materials. I. Hastings, Garth W. II. Ducheyne. Paul. III. Title. IV. Series. [DNLM: 1. Bone and bones--Anatomy and histology. 2. Bone and bones--Physiology. 3. Tendons--Anatomy and histology. 4. Ligaments--Anatomy and histology. 5. Biomechanics. WE 200 P318]
QP88.2.H35 1984 620. 1'98 83-7586
ISBN 0-8493-6264-4

A Library of Congress record exists under LC control number: 83007586

ISBN 13: 978-1-315-89580-2 (hbk)
ISBN 13: 978-1-351-07490-2 (ebk)

Visit the Taylor & Francis Web site at http://www.taylorandfrancis.com and the
CRC Press Web site at http://www.crcpress.com

SERIES PREFACE

Biomaterials science is concerned with surgical implants and medical devices and their interaction with the tissues they contact. Their study, therefore, includes not only the properties of the materials from which they are made, but also those of the tissues which will accept them. Metals, ceramics, and macromolecules are the artifacts. Bone tendons, skin, nerves, and muscles are among the tissues studied. Prosthetic materials, implants, dental materials, dressings, extra corporeal devices, encapsulants, and orthoses are included among the applications.

It is not only the materials *per se* which interest the biomaterials scientist, but also the interactions in vivo, because it is at the interface between implant and tissues that the success of a procedure will be decided. This approach has led to the concept of a more aggressive role for biomaterials in the actual treatment of disease. Macromolecular drug delivery systems are receiving considerable attention, especially those with the capacity for targeting specific sites in the body. Sensing and control of body processes is a logical extension of this. There is much to be done before these newer developments become established.

The science of biomaterials has grown and developed over the last few years to become an accepted discipline of study. It is opportune, therefore, to systematize the study of biomaterials in order to improve their application in medical science, since that is the end point of all studies. That is the aim of this series of books on *Structure-Property Relationships in Biomaterials*. Knowledge of structure and the influence on properties is fundamental to any materials science study; it is a more complex problem to obtain the knowledge from tissue materials, as the living organism has a great capacity for change and adaptation in response to a stimulus. The stimulus may be chemical, electrical, or mechanical. The biomaterials scientist endeavors to identify and to use these stimuli and responses to improve the in vivo acceptability of the materials.

Many institutions and agencies have promoted the science of biomaterials. Societies now exist for this purpose. The Biological Engineering Society (U.K.) founded in 1960 formed a Biomaterials Group in 1974. In the same year the Society for Biomaterials was founded in the U.S. The European Society for Biomaterials (1976) was followed by Canadian and Japanese Societies (1979). All societies play a major role in disseminating knowledge through conferences and publications.

This series is complementary to these society activities. It is hoped that it will not only provide a basis of knowledge, but also its own stimulus for further progress. The series is inevitably selective. In part this is due to the editors' choice, in part to the availability of authors. The editors wish to thank those who fulfilled their agreements. Without them this series would not have been possible.

G. W. Hastings
Series-Editor-in-Chief

PREFACE

NATURAL AND LIVING MATERIALS: THE MUSCULO-SKELETAL SYSTEM

This volume turns from the man-made to the man-making materials. The source for the study of man is man, and the only relevant information for the development of implants and prostheses comes from that same source. The study of tissue materials is of interest in itself; for example, bone may be studied as a complex composite structure without reference to its in vivo condition. The biomaterials scientist is interested in the treatment of disease and injury by the use of new materials and by further understanding of living materials. The interaction, at all levels, of implant materials with the living body is of more interest and importance to patient care than the isolated study of the metals, plastics or ceramics.

That is the context of this volume, and the introduction attempts to underline in more detail some of the reasons for the study of tissue materials. There has been a long-lived controversy over studies in vitro against those in vivo, and this question is addressed in depth in the second chapter. Having set the guidelines, specific aspects of the musculo-skeletal system are considered, the main emphasis being on bone. Structure and different mechanical properties of varying types of bone are discussed by the authors, with the concluding chapter dealing with mechanically mediated electrical properties and their relevance to bone structure and to growth or repair.

No final position has been reached. There is, however, a solid core of information to serve as the basis for future progress, and it has been the aim in this volume to present this for use by all interested in the subject.

The editors wish to thank the authors for their hard work without which no series would have been possible and hope that they and the readers will be encouraged to continue in these areas, so that implant-tissue interface reactions can be studied with the knowledge that the tissues themselves are understood more clearly.

G. W. Hastings and P. Ducheyne

THE EDITORS

Garth W. Hastings, D.Sc., Ph.D., C.Chem., F.R.S.C., is a graduate of the University of Birmingham, England with a B.Sc. in Chemistry (1953) and a Ph.D. (1956) for a thesis on ultrasonic degradation of polymers. After working for the Ministry of Aviation he became Senior Lecturer in Polymer Science at the University of New South Wales, Sydney, Australia (1961 to 1972). During this time he was Visiting Professor at Twente Technological University, Enschede, The Netherlands (1968-69), advising on their program in biomedical engineering. While in Australia, he became associated with Bernard Bloch, F.R.C.S., Orthopedic Surgeon, Sydney Hospital, and began a fruitful collaboration in the uses of plastics materials in surgery.

In 1972 he returned to England as Principal Lecturer in the Biomedical Engineering Unit of the North Staffordshire Polytechnic and the (now) North Staffordshire Health District with responsibility for research. With a particular interest in biomaterials research his own work has encompassed carbon fiber composites for surgical implants, adhesives, bioceramics, prosthesis performance in vivo, and electrical phenomena in bone. He is a member of British and International Standards Committees dealing with surgical implants and of other professional and scientific bodies, including Companion Fellow of the British Orthopaedic Association and Editor of the international Journal *Biomaterials.* He was elected President of the Biological Engineering Society in the U.K. (B.E.S.) in October, 1982. He was awarded a D.Sc. from the University of Birmingham in 1980 for a thesis in the field of biomedical applications of polymers. He has recently been appointed Acting Head of the department.

Paul Ducheyne, Ph.D. obtained the degree of metallurgical engineering from the Katholieke Universiteit Leuven, Belgium, in 1972. Subsequently he worked at the same university towards a Ph.D. on the thesis "Metallic Orthopaedic Implants with a Porous Coating" (1976). He stayed one year at the University of Florida as an International Postdoctoral N.I.H. Fellow and a CRB Honorary Fellow of the Belgian-American Educational Foundation. Thereafter he returned to the Katholieke Universiteit Leuven. There he was a lecturer and a research associate, affiliated with the National Foundation for Scientific Research of Belgium (NFWO). He recently joined the University of Pennsylvania, Philadelphia, as an Associate Professor of Biomedical Engineering and Orthopedic Surgery Research.

Dr. Ducheyne has published in major international journals on mechanical properties and design of prostheses, porous materials, bioglass, hydroxyapatite, and microstructural methods of analysis of biomedical materials. He is member of the editorial board of *Biomaterials, Journal of the Engineering Alumni of the University of Leuven, Journal Biomedical Materials Research,* and *Journal Biomechanics and Comtex System for Biomechanics and Bioengineering.*

He became active in various societies and institutions and has held or is holding the positions of Chairman-Founder of the "Biomedical Engineering and Health Care Group" of the Belgian Engineering Society, Secretary of the European Society for Biomaterials and member of the Board of Directors of Meditek (Belgian Institution to promote biomedical industrial activity).

CONTRIBUTORS

Jonathan Black, Ph.D.
Professor of Research in Orthopaedic Surgery and
Professor of Bioengineering
University of Pennsylvania
Philadelphia, Pennsylvania

W. Bonfield, Ph.D.
Professor of Materials and
Head of Department of Materials
Queen Mary College
London, England

Adele L. Boskey, Ph.D.
Associate Professor of Biochemistry
Cornell University Medical College
Associate Scientist
The Hospital for Special Surgery
New York, New York

Ian W. Forster, MBBS, FRCS, FRCS E
Consultant Orthopedic Surgeon at The General Hospital and
Senior Lecturer in Orthopedic Surgery
Nottingham University
Nottingham, England

J. Lawrence Katz
Professor of Biophysics and Biomedical Engineering
Chairman, Department of Biomedical Engineering
Rensselaer Polytechnic Institute
Troy, New York

Roderic S. Lakes
Associate Professor
College of Engineering
University of Iowa
Iowa City, Iowa

M. Martens, M.D.
Orthopedic Department
University Hospital
Pellenberg, Belgium

K. Piekarski, Ph.D.
Professor
Mechanical Engineering Department
University of Waterloo
Waterloo, Ontario

Aaron S. Posner, Ph.D.
Professor of Biochemistry, Cornell University Medical College
Associate Director of Research, Hospital for Special Surgery
New York, New York

R. Van Audekercke, Ph.D.
ICOBI, Biomechanics Section
Heverlee, Belgium

TABLE OF CONTENTS

Chapter 1

INTRODUCTION — THE STUDY OF TISSUE MATERIALS

G. W. Hastings and P. Ducheyne

TABLE OF CONTENTS

I. THE STUDY OF TISSUE MATERIALS

The subject of biomaterials includes not only materials of synthetic origin intended for implantation in the human body in order to support or replace some of its structures but includes also those same bodily structures, i.e. the tissue materials. In its wider context the subject of biomaterials should include all those derived from living organisms but is generally restricted to those found in the human body. This is because the aim of biomaterials research is to assist in the medical care of the human patient, and hence biomaterials research embraces all pertaining to that aim. The particular study of human tissue materials will contribute to an understanding of the fundamental properties of those tissues, to the establishment of relationships between normal and pathological conditions, and to the study of interactions between tissues and surgical implants. The study of human tissue does not deny the value and importance of studying other biological materials, and considerable information may be gained which may be of direct reference to human studies.

A conference on the theme of mechanical properties of biological materials led to a publication which provides a useful review of the whole subject.[1] In that book, Gordon[2] reminds us that the study of biomechanics requires a level of understanding well beyond that of elementary texts, and in particular, since the magnitude of mechanical strains observed in biological systems are beyond those normally considered in the usual engineering materials, a new design concept is required. This implies that the biomaterials designer should consider the exceptional functional adaptation of the materials in living organisms. The relationship between strength and weight and the use of composite materials and structures in vivo should direct the attention away from the more massive designs common to ordinary engineering practice. There is, furthermore, a more dynamic interaction between materials and environment than would usually be considered, and this is expressed in the principle of homeostasis and cellular activity. Living materials respond not only to external chemical control processes, but also to mechanically or electrically mediated effects. The well-known response of the skeleton to applied external force or to internal muscle action, commonly expressed asWolff's law, illustrates this. Similarly, the gross chemical analysis of biological materials cannot define the role of the elements so identified. It is their position in a structure and their movement in and out of it that is the critical factor.

The point to be made is that simplistic studies of biological systems or materials may be very misleading. Dynamic interaction is occurring between all components to maintain the homeostasis within very narrow limits. Similarly, the range of parameters determining biological acceptance of a synthetic biomaterial may be operative over closely defined limits and may be only adequately understood, and thereby controlled, if a detailed study is made of the tissue materials as well.

A study of biomaterials which does not include the study of living materials will be defective and may be dangerously misleading.

II. BIOLOGICAL ACCEPTABILITY

Biological acceptability often referred to as biocompatibility is the study of the interaction of implant materials with the tissues which receive them. Inertness is a concept not found in implant acceptance. Every foreign material and even one taken from a donor site in the same patient, results in trauma to the tissues receiving it. It is the level of that reaction which permits classification of a material as acceptable biocompatible or otherwise. Surgical catgut produces a reaction, but its usefulness has been demonstrated successfully in vivo over many years of experience and the reaction is acceptable. Biological acceptability is often demonstrated retrospectively. Ultra high molecular weight polyethylene is known to be bio-acceptable because there is an accumulated fund of knowledge about its in vivo performance in joint replacements extending over several hundred thousand cases. It is not tolerated in vivo under all conditions. Direct articulation with cartilage is undesirable and leads to necrosis of adjacent tissue. Large volumes of wear debris are produced which may be the cause of cell death.

Biological acceptability should be examined at various levels of interaction:

1. the interaction of bulk material and tissues, including chemical reactions, shape factors or density effects;
2. the reaction of debris resulting from wear;
3. the effect of corrosion or degradation products;
4. mechanical factors — modulus of elasticity imbalance, movement.

All of these imply a two way interaction, implant on tissues and tissues on implant. All require knowledge of in vivo conditions at the site of use and hence of the properties of the tissue biomaterials themselves.

Although any program of biomaterials development could not proceed without in vitro evaluation of materials, it is now understood that laboratory wear tests, simulator studies, and tissue culture reactions provide only a partial satisfaction for the questions posed. They are useful and necessary to isolate single parameters, i.e., to reduce the number of variables operating and to provide a systematic framework for evaluation. Eventually, there must come a stage when only human trials will provide the final answer. Furthermore, those trials need to be long-term, and no material or system can be given universal acceptance as a result of short-term trials in a very limited number of patients obtained by careful selection. In fact, one of the pressing needs in biomaterials research is for more long-term evaluation to be carried out. It is costly in terms of laboratory space and experimental requirements but, in terms of patient reliability of implants, is of immense value. The examination of retrieved implants is an area for study which yields information from the only valid test of human experience. Results are now available from 20 years of human hip joint replacement, and it is vital that valuable information should not be lost. This again serves to emphasize the need to understand as far as possible the biological and biomechanical principles underlying host tissue-implant responses.

One might even go further to include patient response, i.e., psychological and social factors and the provision of complete treatment planning and patient instruction. The reason is that materials should not be studied in isolation from the environment in which they are to be used, and this includes all relevant factors. The way a patient responds to a replacement part and the lifestyle of that patient are very relevant to implant performance, since a new situation has been artificially produced in which patient plus implant interact with the external environment.[3]

III. INFORMATION TO BE OBTAINED

Information obtained from the study of living materials includes:

1. a developing fund of knowledge of tissue properties and structure;
2. formation of a basis for the understanding of tissue-implant reactions;
3. a definition of parameters required adequately to treat disease and injury;
4. an examination of genetic factors and relevance of the external environment, i.e., epidemiological factors in disease or injury;
5. work-study programs to improve conditions of employment not only for the general population but for injured or permanently handicapped people (i.e., ergonomics) may possibly be considered a study of living human material in actual life.

To illustrate this, the pioneering work of Pauwels[4] on the treatment of osteoarthrosis of the hip showed how remodeling of the proximal femur and the acetabular region of the pelvis takes place when the forces at the hip joint are redirected. Thus, not only is the response of bone to various forces better understood, but more effective planning for osteotomy treatment of this condition can be made. In time this leads to a more careful consideration of the indications for total joint replacement. Similarly, McKibbin[5] has reviewed the various mechanisms by which fractures heal, and systematic work into the histology of fracture healing lays the framework for understanding the mechanisms and therefore for treatment planning.

The study of strain-generated potentials in bone is another example of the examination of materials to determine the structural origin of an observed phenomenon. As a result, it now appears that these potentials are primarily a consequence of piezo-electric behavior in response to an applied strain. In conjunction with these investigations, a clinical method has emerged for the treatment of fracture non-unions. An electrical or magnetic field, the former applied by implanted DC electrodes, the latter by external coils, appears to stimulate bone healing. As part of the scientific investigation accompanying the clinical studies, cell culture methods have been used to investigate which in vitro characteristics of the field are important. The way in which cells transduce mechanical forces into electrical energy may help to elucidate the control systems operating in vivo. If the acceptable limits of strain permissible at a fracture site during healing can be defined, this will lead to better design parameters for fixation materials, devices, and methods.

IV. THE AIM OF THIS VOLUME

In the previous volumes of this series, the various synthetic biomaterials, metals, ceramics, and polymers have been discussed. It is, therefore, the aim of this volume to present some of the details necessary in order to define and assess tissue material problems. The chapters should provide a basic understanding for the study of human skeletal tissue materials and at the same time give an account of the present state of the art for some of the more important areas of development. Of necessity, there has been selection of topics, but is hoped that sufficient is given to meet the needs of those involved in this type of biomaterials research and application.

An understanding of the structure and properties of biomaterials in living systems should lead ultimately to improved synthetic materials with more complete biological compatibility (i.e., satisfying biological biochemical and mechanical requirements). This in turn will lead to improved treatment offered to patients.

REFERENCES

1. **Vincent, J. F. V. and Curry, J. D., Eds.,** *The Mechanical Properties of Biological Materials*. Soc. for Exp. Biol. Symp. 24, 1980. Cambridge University Press, Cambridge, 1980.
2. **Gordon, J. R.,** Biomechanics: the last stronghold of vitalism, in *The Mechanical Properties of Biological Materials*, Vincent, J. F. V. and Curry, J. D., Eds., Cambridge University Press, Cambridge, 1, 1980.
3. **Jansons, H. A.,** Artificial bioconstructions as behavioural systems. *Biomaterials*, 3, 52, 1982.
4. **Pauwels, F.,** *Biomechanics of the normal and diseased hip*, Springer Verlag, Berlin, 1976.
5. **McKibbin, B.,** The Biology of fracture healing in long bones. *J. Bone Jt. Surg.* 60B, 150, 1978.

Chapter 2

TISSUE PROPERTIES: RELATIONSHIP OF IN VITRO STUDIES TO IN VIVO BEHAVIOR

Jonathan Black

TABLE OF CONTENTS

I. THE FLOPPY-EARED RABBITS

Any approach to the question of the differences between tissue mechanics in vivo and in vitro is usually greeted with considerable resistance and a certain surprise that any question of differences should be raised. In this chapter, I propose to examine the evidence that is available to relate properties in vitro with those in vivo. In addition I shall, gently I hope, attempt to point out problems in many reports and their interpretations that have led to the sorry state of the question today.

For it is safe to say that, in the first place, there are profound differences between properties of tissue *in situ* and in the laboratory, and furthermore, we have very little solid data to permit us to predict one from the other.

How have we arrived at this state? Preparation of this chapter has in many ways resembled a detective story. However, at the onset I was reminded of the story of the Floppy-Eared Rabbits. It is worthwhile taking a few moments to review this strange sidelight to biological research and to determine if a relevant lesson may be extracted.

Barber and Fox[1] discussed this story and first brought it to light. It originated in work reported by Lewis Thomas[2] in 1956 in conjunction with another, somewhat unrelated research effort.

Thomas performed the following experiment: papain was obtained as a crude extract of natural latex. It was found that 1 to 10 mg injected into 800 to 1000 gr rabbits brought about rapid ear collapse, starting about 3 hr post injection and completed by 48 hr. Arterial ligation (of the ear) for 15 min after injection prevented collapse of the ear, in the presence of a 10 mg injection. Significant histological changes were seen in the ear by 48 hr after injection, and ''recovery'' occurred over a period of several days thereafter. Cortisone effectively blocked ear ''recovery'', and there were signs of a progressive immunity (failure to collapse) with repeated injections of the papain preparation.

By itself, this report should be of interest here as it suggested profound mechanical property changes, in vivo, in response to very subtle changes in the internal environment. Thus, we should not be surprised if we encountered equally significant post-mortem or post-excision changes in mechanical properties.

However, the interest in this report lies in the report by Barber and Fox that Thomas' observations of ear collapse were not the first discovery of this effect. It is clear that this occurred in a study performed earlier by Kellner et al.[3] and reported in 1951. That is, Kellner used a similar preparation and injection scheme in an attempt to produce arteriosclerotic lesions in rabbits. Although it is clear that Kellner must have observed both the collapse and recovery phenomena, he did not report it in the original paper. According to Barber and Fox, although Kellner routinely used the effect as an assay for papain potency, he did not report it because of '' . . . his research preconceptions and the occurrence of other serendipitous phenomena in the same experimental situation.''

Barber and Fox call this process ''retrospective falsification'' — that is, the deliberate overlooking and non-reporting of unforeseen developments, by lucky or happy chance, that are not central to the investigation.

In this chapter, I suggest as a central thesis that such retrospective falsification, tidied up and called ''biological variability'', ''preconditioning'', etc., has acted for a long time to obscure the real evidence for the relationship between tissue mechanics in vivo and in vitro. Furthermore, it appears that failure either to consult previous art or to evaluate it critically has lead to practices that obscure post-mortem tissue changes.

II. EVIDENCE FOR LACK OF PROPERTY CHANGES AFTER TISSUE EXCISION

It is a common practice in tissue mechanics to preserve tissue by freezing before testing.

Table 1
SUMMARY OF SEDLIN AND HIRSCH'S[4] EXPERIMENTS ON ENVIRONMENTAL EFFECTS ON PHYSICAL PROPERTIES OF CORTICAL BONE

Question	Material	Test type	Conditions		Results
Effect of temperature	77S (3 sub)	Bending(3P)	1 37°	1 − 2:	5% ↑ M (1 sub)
			2 21°		6% ↑ d_u
Effect of air drying	137 S (5 sub)	Bending(3P)	1 Ringers	6 − 1:	20% ↑ σ_u
			2 5 min AD		→ M
			3 10 min AD	7 − 1.	→ σ_u
			4 15 min AD		sig ↑ M
			5 30 min AD		sig ↓ E_u
			6 60 min AD		sig ↓ d_u
			7 1 wk 105°C		
Effect of prior freezing	74S (3 sub)	Bending(3P)	1 3 hr RT	2 − 1:	9.5% ↓ σ_u
			2 3/4 wk −20°C		→ M
					→ E_u
					→ d_u
Effect of fixation	15S (1 sub)	Tension	1 RT	2 − 1:	→ M
			2 3 wk 10% formalin		

Key: S = specimens; sub = subjects; AD = air drying; RT = room temperature; 3P = three point; M = modulus; d_u = deflection to failure (bending (3P)); E_u = energy to failure; σ_u = ultimate stress; ↑ = increase; → = unchanged; ↓ = decrease; sig = significant change, amount unreported

This has come about through expediency in that it is usually difficult to coordinate sacrifice of test animals or excision of tissue from patients with the timing of the planned research protocol. The justification for this practice, in most cases, is a reference to the work of Sedlin and Hirsch.[4] This paper entitled, "Factors Affecting the Determination of the Physical Properties of Femoral Cortical Bone", appears to be the standard reference on this subject. It is a large, complex group of studies that investigated the effects of temperature, air-drying, freezing, and fixation as well as intrinsic conditions such as porosity, mineralization, etc. on the mechanical behavior of bone. In Table 1, I have summarized the experimental material, test-type, conditions and results of experiments of the four groups of data that address the aspects of tissue condition.

The group of experiments, on which is based the conclusion that freezing at −20°C is a satisfactory method of preservation of tissue, is shown in the third section of this table. Seventy four specimens were obtained from three human cadavers and tested in three-point bending. Two groups of specimens were used. One group was tested 3 hr after excision, and storage at room temperature. The other group was tested after freezing, storage at −20°C for 3 to 4 weeks, and then defrosting to room temperature. The conclusions that Sedlin and Hirsch drew from this experiment are as follows: they found a 9.5% *decrease* in ultimate stress when comparing the frozen to the unfrozen group, while they found no significant change in the modulus, the energy to failure, or the deflection to failure.

It should be immediately obvious to the reader that this is a static, concentrated force experiment. Thus, the conclusions that are drawn really address only static experiments that progress to failure. Furthermore, the finding of a substantial (9.5%) decrease in ultimate stress should be a warning that the properties of the tissue are not totally unchanged by the period of cold storage and the associated freezing-defrosting cycle. However, let me draw the reader's attention to a still more serious criticism of this experiment. The authors state that the source of the tissue used in this experiment was the midfemoral diaphysis of adult subjects, that were removed at routine post-mortem examination on either the day of death or the day *following* death. Thus, this experiment, as reported, does not investigate the effect of death on mechanical properties of tissue. What it does investigate, in a very limited way, is the ability of freezing, with a single freeze-thaw cycle, to preserve certain ultimate properties in bending of bone which has been dead, in the sense of not participating in systemic physiology, for periods of 8 to 32 hr. It should be clear that the other results reported in Table 1 must be interpreted with a similar caveat.

III. THE CURRENT VIEW ON THE RELATIONSHIP OF IN VITRO MEASUREMENTS TO IN VIVO PROPERTIES

Evans,[5] in his major review, *Mechanical Properties of Bone*, devotes an entire chapter (Chapter 3) to a thorough discussion of the properties of living vs. dead bone. His analysis depends upon a review of four studies: Stevens and Ray,[6] Lissner and Roberts,[7] Greenberg et al.,[8] and Gurdjian and Lissner.[9]

Stevens and Ray[6] conducted an experiment in which they attempted to determine whether there were differences in properties of bone in rats dependent upon the state of the bone viability. The fundamental experiment consisted of amputating the right hind leg of male rats, excising the tibia, defleshing, weighing, and then devitalizing the bone by exposure to an acetone-carbon dioxide snow mixture for 5 min, followed by 37°C exposure for 5 min; the entire cycle was repeated three times. Then the (assumed) dead tibia was reweighed and implanted within the sheet of the rectus abdominus muscle of the same animal from which it had been taken. Groups of six animals were sacrificed at periods of 3 hr to 120 days after implantation of the dead tibia. At sacrifice, the tibias were removed, both the dead one from the rectus abdominus and the live tibia from the intact left hind leg and a number of studies were conducted:

1. Whole bone weight.
2. Radiographic comparison.
3. Whole bone strength.
4. Specific gravity of the cortex.
5. Ash weight of the bone.

The mechanical testing was conducted by a dead weight, three point bending technique. Overall, Stevens and Ray found little difference in the breaking strength of the right vs. the left tibiae, when compared on a group mean basis. However, several points attract our attention.

Figure 1 (adapted from Stevens and Ray[6]) shows the stress-strain or more properly load-deformation curve for a pair of bones removed 45 days after insertion of the devitalized bone. Comparison of the live to the dead one displays the following results:

1. The apparent load to failure of the live bone exceeds that of the dead bone by 42%.
2. The stiffness of the live bone is approximately 120% of that of the dead bone.
3. The shape of the stress-strain curves differ radically: the live bone displaying an almost linear behavior to failure, while the dead bone displays a distinct ''upcurl'' in the last 7 to 8% of load application before failure.

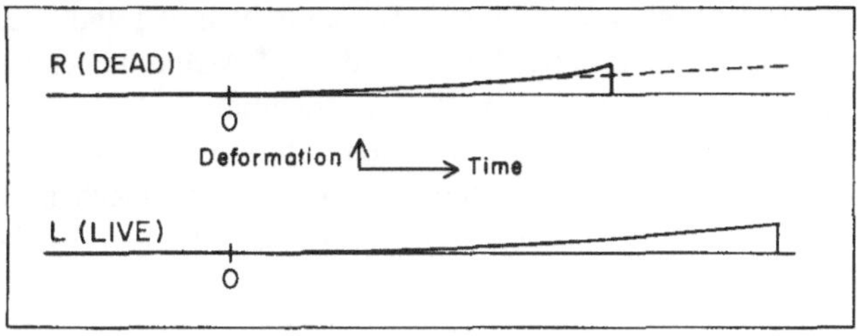

FIGURE 1. Load-deformation curves for live and dead bone. (Adapted from Stevens and Ray.[6])

Stevens and Ray, (Reference 6, p. 417) summarize the differences as follows: "Towards the point of failure, however, deformation of the dead tibia increases rapidly, indicating that the elastic limit has been passed. The living tibia retains its elastic properties until the point of failure. There was little difference between the tracings of the living and dead bone from those groups of animals killed soon after operation."

At a later point in the same paper (p. 422), they state: "Some physical properties of living and dead bone have been studied in rats; most of these are interrelated and ultimately depend upon the composition of the tissue."

Evans[5] provides an excellent discussion of the details of this experiment. However, his conclusion is indicated in the following quotation (p. 38): "The fact that Stevens and Ray found no statistically significant differences among the mechanical properties of living and dead bone is not surprising when one considers the bone is primarily composed of non-living material . . . "

I suggest that both Evans and Stevens and Ray missed the significant part of this study. That is, although there are not gross differences in the mean values of the strength parameters of bone that were measured, there is clearly a *qualitative* difference between dead tissue stored in vivo for 45 days and live tissue. It would be interesting to know if these qualitative differences reported by Stevens and Ray were exhibited throughout the time course of their study.

The study of Lissner and Roberts[7] consisted of comparing strains produced in several of the vertebrae of living anesthetized and dead embalmed dogs under controlled acceleration impact. The recording of strain was made possible by the implantation of strain gauges directly on processes of the lumbar and thoracic vertebrae of these animals. Data presented and discussed by Evans consist primarily of the relationship between the dynamic load factor which is the ratio between strain at a constant acceleration and the initial value obtained with a particular change in acceleration and jerk which is the rate of change of acceleration. Differences were found in these parameters at several of the vertebral locations studied, but they amounted to perhaps 20% which were felt to be within the maximum variation of the strain gauge technique being used. Furthermore, it was difficult to ascribe the differences to a comparison of the live and dead conditions, since the dead animals had, necessarily, been embalmed. Greenberg et al.[8] performed a similar study which was also discussed by Evans. However, in this experiment, measurements were conducted under an additional condition, that is, on the dead animal immediately after sacrifice but before embalming. In addition, some comparisons were made in breaking loads and stiffnesses of the tibiae in vivo, freshly excised and after one week's embalming. Results here are similar to those reported by Lissner and Roberts[7] and earlier results by McElhaney et al.[10] who examined the effect of embalming on stiffness and strength of bovine bone. That is, very small changes were seen between the three cases, again with the evaluation being conducted in terms of quantitative parameters rather than qualitative behavior.

The final study referred to by Evans is that of Gurdjian and Lissner[9] who used a stress coat technique to study the pattern of strain in the skulls of living and dead monkeys and the necessary comparison with the same skull after drying. Again the conclusion is similar; that is, that the patterns are very similar in all three conditions.

From an evaluation of this literature, Evans is led to this conclusion: (p. 42): "Contrary to what may have been expected, the differences between the mechanical properties of biomechanical behavior of living and dead bone, when tested under similar experimental conditions, are few. Consequently, results obtained from experiments with dead bone, especially in the fresh or moist condition, appear to be valid for extrapolation of comparable properties and behavior of bone in the living animal."

In a more recent review, Wall et al[11] examined many of the same aspects that Evans reviewed and concluded, with respect to effects of time between death of the donor and removal of the tissue: (p. 176) " . . . [the] interval between death and removal may have some bearing on the results and should be noted". They further emphasized the need to test fresh material and suggested that prolonged freezing could alter the mechanical properties of bone. This is in agreement with the results (but not the conclusions) of Sedlin and Hirsch.[4] The mechanism suggested by Wall et al.[11] is disruption of collagen fibers by the formation of internal ice crystals. They also suggested that freezing bone in saline solution could result in an unanticipated dehydration due to the formation of ice crystals of different sizes within and outside the specimen.

It seems clear that the conventional wisdom that freezing does not affect tissue properties and that life-death transitions play a negligible role in determining tissue properties in vitro must be severely questioned. I propose now to examine some of the better studies that have attempted to relate measurements in vitro with properties of musculo-skeletal tissues in vivo.

IV. EFFECTS OF POST-MORTEM CHANGES ON TISSUE MECHANICS

A. Tendons and Ligaments

One of the pioneer investigators in the properties of tendons and ligaments, A. Viidik, was concerned with the effect of postmortem storage on the behavior of these materials. Two papers published in the 1960s reported early experiments in which he attempted to investigate these problems.

The first paper (Viidik et al.[12]) was a comparison of the failure in tension of the anterior cruciate ligament in the knee of the rabbit. The two groups that were studied were a group at sacrifice and a group in which the animal was killed and the joint kept closed for six hr before testing. Viidik found a significant increase in the rate of internal failures in the ligament, prior to total failure, as seen by dips in the load-elongation curve at six hr compared to the immediate postoperative, i.e., postsacrifice testing. The other parameters studied (modulus, elongation of failure, force to failure, and energy to failure) were apparently unchanged when comparing the two groups.

Using this earlier study as a control group, Viidik and Lewin[13] then went on to compare several common methods of storage after sacrifice. The four test groups that were used are

1. Storage in 0.9% saline at 20°C for 5 hr postmortem.
2. Storage in 0.9% saline at +4°C for 24 hr postmortem.
3. Storage of the intact joint at −20°C for 1 week postmortem followed by a rapid thaw in water and
4. Partial dissection of the joint and storage in 10% formalin for 6 days.

Table 1 from their paper is reproduced here with some modifications as Table 2. It shows an irregular pattern of differences distributed among these four groups of eight animals each when compared to the initial control group from the previous study. The parameters that were evaluated are energy to failure, load to failure and elongation to failure.

Table 2
PROPERTIES OF RABBIT ANTERIOR CRUCIATE LIGAMENTS AS A FUNCTION OF POST-EXCISION STORAGE CONDITIONS

Mean values ± standard error

Grp	n	tan α (°)	Failure energy (kpmm)	Failure load (kp)	Elongation at failure (mm)	Body weight (kg)
N	14	1.53 ± 0.07	36.0 ± 5.9	26.9 ± 1.1	1.74 ± 0.07	3.04
A	8	1.73 ± 0.21	70.8 ± 7.3* ↑	40.3 ± 2.8* ↑	2.94 ± 0.31* ↑	3.36
B	8	1.18 ± 0.08* ↓	46.0 ± 4.0	27.4 ± 2.6	2.11 ± 0.22	3.41
C	8	1.81 ± 0.15	53.2 ± 4.9* ↑	33.2 ± 3.0	1.71 ± 0.17	2.75
D	8	1.06 ± 0.12* ↓	23.9 ± 3.1	19.5 ± 1.9* ↓	1.56 ± 0.14	2.91

Note: * denotes a significant difference ($p < 0.05$) from the controls (N), ↑ increase, ↓ decrease.

After Viidik and Lewin.[13]

Furthermore, qualitative differences, although not statistically significant, were seen in the manner in which the ligaments failed. This is reminiscent of the study of Stevens and Ray,[6] in that in the presence of small quantitative changes, qualitative changes in the experiment were observed.

Viidik[13] arrived at the following conclusions (p. 153): "The conclusion from this investigation is that mechanical tests of collagenous tissue removed in surgery and without protection of the surrounding tissue must be done immediately. If this is not feasible, some adequate method of storage must be developed. Possibly some liquid like blood, plasma, or synovial fluid providing low exchange gradients of different molecules can be used.

"When carrying out histological investigations, the material must be fixed immediately after removal at surgery if the evaluation of finer cellular structure, stainability, etc. is essential."

Tkaczuk[14] performed a study of human lumbar longitudinal ligaments utilizing similar techniques. His material, like that of Sedlin and Hirsch,[4] was obtained at autopsy and tested "within 48 hours of death" (p. 19). He concluded that most changes were due to a loss of water from the tissue during testing. Rapid freezing (to −60°C) and thawing failed to show any changes in elongation (to 500 gr load), residual deformation, or energy dissipation per cycle. Despite considerable efforts to maintain water content, a marked (about 50%) decrease in energy dissipation per cycle (hysteresis) occurred uniformly in his studies between cycles 1 and 2, with cycle 3 being essentially identical to cycle 2.

It is possible that some of these changes described by Viidik and by Tkaczuk are related directly to time after death. However, ligaments are relatively inaccessible and require considerable dissection to free them from surrounding tissues. In a similar tissue, tendon, Matthews and Ellis[15] attempted to overcome these problems and determine if changes occurred in properties related to time after death (independent of water loss). They used feline extensor communis and extensor lateralis tendons. The overall stiffness (average apparent elastic modulus) was measured in dynamic extension with the tissue maintained in a stream of Ringer's solution for periods of up to three hr after excision. In addition, a group was quick frozen and stored at −10°C for two weeks before thawing and testing. Although little change was seen in the modulus with time, there is an apparent pattern of increasing modulus with time for higher initial values. In addition, the frozen and stored group showed an average decrease in modulus of 10.2% (significant at the 99% confidence level). What is even more interesting is the observation of a progressive length change (shortening) during

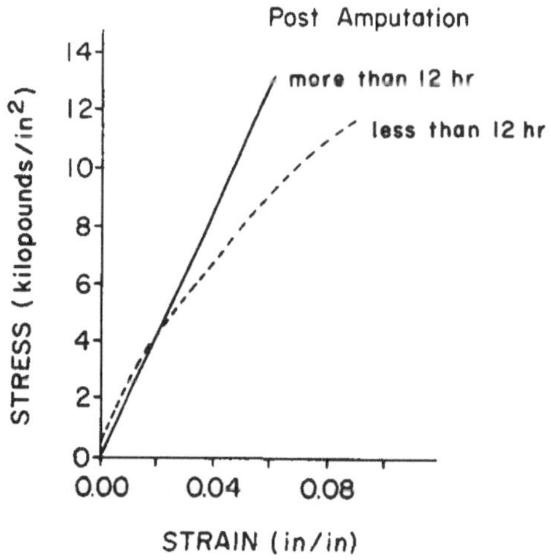

FIGURE 2. Stress-strain curves for unembalmed human extensor tendons. (Adapted from Benedict et al.[16])

the three hr test period, of between 0.1 and 0.7%, despite the fully hydrated condition of the test.

Benedict et al.[16] became involved in this question also in an attempt to elucidate the stress-strain characteristics and strength of fresh human tendons. They obtained their material from amputations so they were able to study a variety of tendons at periods of every four hr after death. In their analysis, they found it necessary to distinguish between tendons tested within 12 hr of amputation and those tested after this period had elapsed, despite the use of refrigerated storage for the limbs. The reason for this can be seen graphically in Figure 2. Here we see the statistical averages for extensor tendons from these two groups in their study. The contrast between the non-linearity of the fresher group and the almost brittle behavior of the less fresh group is striking. Note, however, that while the modulus at higher strains increases with time, the modulus at lower strains (below 1%) actually decreases with time. In this study, flexor tendons showed similar but inverse effects: modulus at higher strains decreased with time, while that at lower strains increased with time.

Both Matthews'[15] and Benedict's[16] studies are less than totally definitive since the initial values were not determined in situ. To my knowledge, such an experiment has not been done for either tendons or ligaments. However, a rather elegant experiment has been performed on feline mesentery membrane.

Chu, et al.[17] have studied the mechanical behavior of feline mesentery tissue in a series of experiments designed to compare in vivo and in vitro behavior. The mesentery is not strictly a musculo-skeletal tissue, being a fascia that supports the intestines. However, it is a two-dimensional, non-excitable tissue composed of a membrane, containing collagen and elastin, supporting two layers of mesothelial cells. It is also reasonably vascular and contains clusters of Pacinian corpuscles, which are the encapsulated ends of sensory nerve fibers.

Although this tissue is in a state of biaxial stress in vivo, Chu tested it in uniaxial tension for convenience. A clamp system was devised that could be used both in vivo and on excised tissue.

Characteristic force-elongation diagrams are seen in Figure 3. The experiment is performed by applying strain increments, with 45 sec permitted for stress relaxation between increments, yielding a saw-toothed "stress-strain" curve. The two experiments on the left portion of the figure are for tissue in situ (or VIS)* with a five minute recovery period between cycles.

* See section VI for discussion of tissue test states and abbreviations.

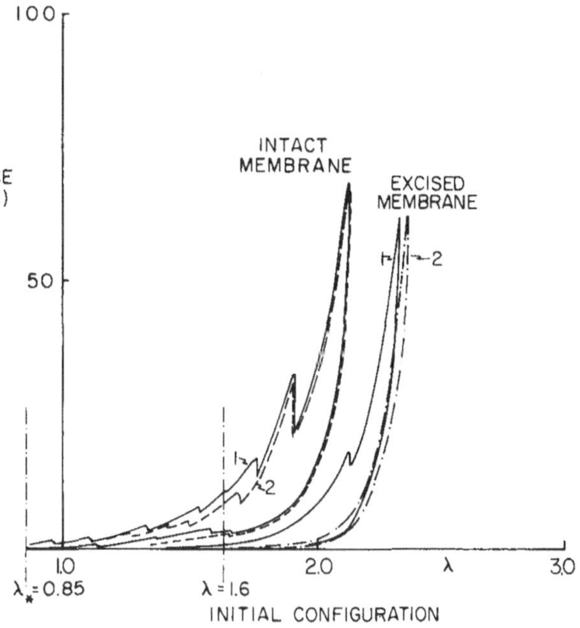

FIGURE 3. Force-elongation curves for feline mesentery membrane. (Adapted from Chu et al.[17])

Following these two tests, the tissue was left in the clamps, allowed to relax completely and then excised. Two additional tests were then performed, following the same protocol. These later conditions, with the tissue in a physiological bath, are properly nonviable, physiologic (NVP).

A dramatic difference may be seen between cycles 1 and 2 on VIS tissue compared with cycles 1 and 2 on NVP tissue. Compared with VIS, NVP conditions produce a considerably softer tissue. In addition, a great preconditioning effect is seen in the changes between the first and second NVP test. In particular, the large hysteresis (energy dissipation per cycle) seen in the first three cycles (including VIS) is nearly absent in this last test.

Chu suggested that the following significant changes occur on tissue excision:

1. The mechanical tethering of the tissue, and any resultant prestress, is lost.
2. A considerable amount of blood is lost from the tissue.
3. All innervation is lost from the tissue.
4. Tissue perfusion is lost.
5. Influx and efflux of tissue metabolites is changed.

One might argue that tissue properties secondary to blood flow and tethering effects are extrinsic behavior. However, this tissue response is the actual in vivo response. Attempts to interpret the mechanical response of the tissue from the second (preconditioned) excised cycle would produce a picture far removed from actual in vivo behavior.

These results are remarkably similar to those previously reported for a quite different preparation.[18] As part of an experiment to investigate the mechanical consequences of postoperative scarring in flexor tendon repair, we developed a central flexor tendon model in the hind paw of the rat. In Figure 4, we see the effect of tendon excursion (of the deep central flexor) after scission of all other tendons and amputation of the rat paw. While this experiment is not done in vivo, the preparation is performed on an anesthesized animal, and the paw is amputated immediately before test. Thus, this represents room temperature or NVM conditions.

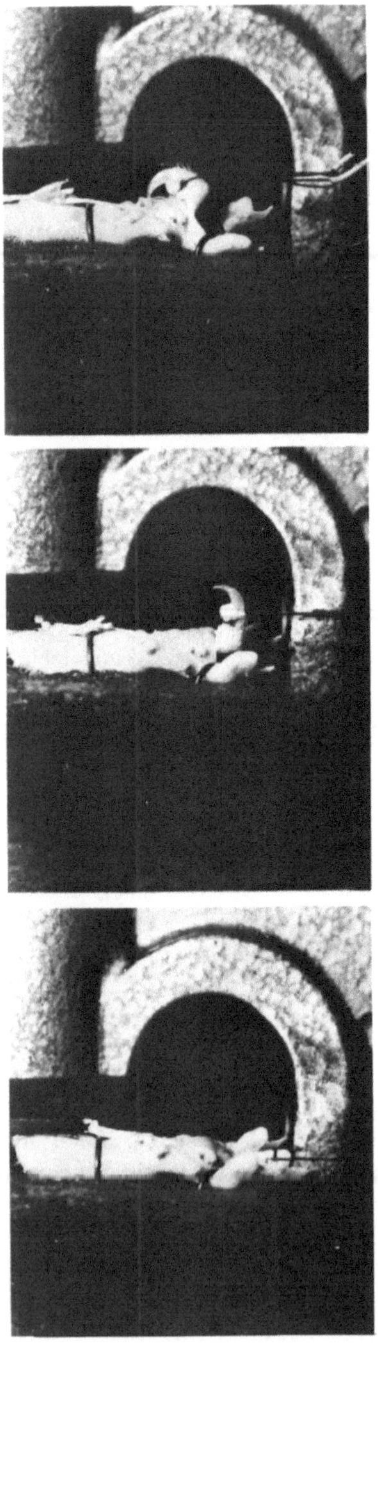

Full extension — 90° Flexion — Full flexion

FIGURE 4. Force-linear excursion behavior in the rat central flexor tendon (hind). (From Lane, J. M., Black, J., and Bora, F. W., Gliding function following flexor-tendon injury, *J. Bone J. Surg.*, 58A, 985, 1976. With permission.)

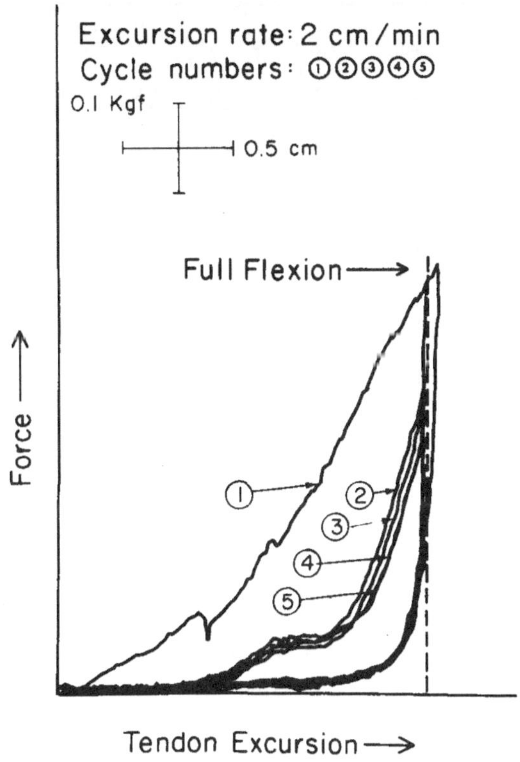

FIGURE 5. Force-excursion traces rat central flexor tendon (hind) control. (From Black, J., Dead or alive: the problem of *in vitro* tissue mechanics, *J. Biomed. Mater. Res.*, 10, 377, 1976. With permission.)

In Figure 5, we see the results of five successive excursions to full flexion. The changes between Cycle 1 and Cycles 2 to 5 are not affected by resting period between cycles. Comparison of these results with those of Chu et al.[17] for excised tissue shows the same pre-conditioning effect and loss of hysteresis after the first cycle.

It should be emphasized that loss of hysteresis in soft tissues occurs upon excision but may also occur in vivo. Barbanel et al.[19] demonstrated this effect in studying the torsion deformation of human skin in situ. Figure 6 is adapted from Figure 7 of their report. Here we clearly see the differences in the first trace when compared to the second and fourth ones. The fifth trace is a reversal of the initial one and is a mirror image of the initial deformation (torque) curve. The authors suggest that there is an initial "slack" in the collagen fibers in the skin. Cycle 1 elongates and orients these fibers and produces the reproducible behavior of cycles 2 to 4. The reversal of Cycle 5 then moves the fibers into a different configuration, with a result much like the initial displacement behavior of Cycle 1. This mechanism had previously been suggested by one of these workers[20] in an effort to explain the effect. Unfortunately no histological data are available to support this assertion. Presumably a longer resting period than the one min between cycles would restore the initial pattern on subsequent cycles.

It is also clear that non-mineralized, high water content tissues, like mesentery, fascia, and skin, can show extremes of post-excision behavior. Perhaps the best illustration of this is the study of Marangoni et al.[21] on the tensile properties of guinea pig skin. As part of a program to study return of strength during wound healing, these investigators sought to determine if the delay involved in completing the excision of 15 to 20 specimens to be

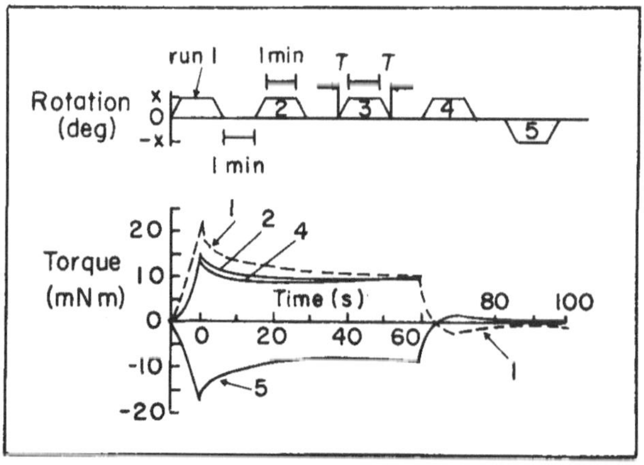

FIGURE 6. Rotation-torque relations in human skin in vivo. (Adapted from Barbanel et al.[19])

obtained from each animal would produce differences between the first and last tested. A control program, utilizing testing at excision and after periods of 1 to 6 hr of storage under three conditions was carried out. The three storage conditions were described as follows: "Sections of the excised tissue were stored in a constant temperature chamber at 1.6°C and 100% relative humidity, 37.8°C and 100% relative humidity or rapidly frozen with liquid nitrogen −224°C and stored in a carbon dioxide cold storage chamber at −92.8°C" (p. 444).

All specimens were equilibrated in a water bath at 24°C before testing. The strips were tested to failure, in tension, at a low strain rate, and four parameters were extracted from the results: maximum stress (ultimate stress), maximum strain (ultimate strain), maximum work input (specific energy of failure), and maximum stiffness.

The results show a complex and highly significant pattern of property dependence on time after excision (Table 3). In general, the quick freezing, although abolishing temporal changes, produced values between 1 and 20% different from immediate excision values. The two non-frozen storage conditions produced the following results: maximum strain and maximum work input both decline sharply with time, while maximum stress and maximum stiffness both rise sharply, reaching a peak about 30 min after excision, and then decline with time.

Since storage at a nominal 37°C is the usual treatment for tissue when one attempts to maintain VIV conditions, the full, normalized results for the 37.8°C storage are shown in Figure 7.

This study makes it clear that both time after excision and method of storage can affect the properties of soft tissues in a significant manner; changing both qualitative and quantitative aspects of mechanical behavior. As a generalization, these changes agree with the observation of Smith[22] in his study of rabbit anterior cruciate ligaments:

"In my own experience, ligaments become less readily extensible (Young's modulus increases) and react elastically to greater loads (the elastic limit increases): the change becomes apparent within an hour of death and is progressive thereafter." (p. 371).

B. Cartilage

Articular and costal cartilages are highly organized tissues with water contents approaching 75 and 60% respectively. The mechanical properties of this tissue type are a reflection of both deformation of the solid matrix as well as fluid flow as a response to strain gradients. Thus, its mechanical properties are extremely sensitive to changes in association between water and matrix.

Table 3
PROPERTIES OF GUINEA PIG SKIN AS A FUNCTION OF STORAGE TEMPERATURE AND TIME AFTER EXCISION

		Percent Change from Immediate Post-Excision			
Time after excision	Storage temperature (°C)	Maximum strain (%)	Maximum work input (%)	Maximum stress (%)	Maximum stiffness (%)
3 hr	−92.8	−1	−5	+7	+20
	+1.6	−26	−44	−3	+45
	+37.8	38	19	+35	+160
6 hr	−92.8	−1	−5	+7	+20
	+1.6	−30	−65	−31	+30
	+37.8	−40	−69	−7	+120

Note: After Marangoni et al.[21]

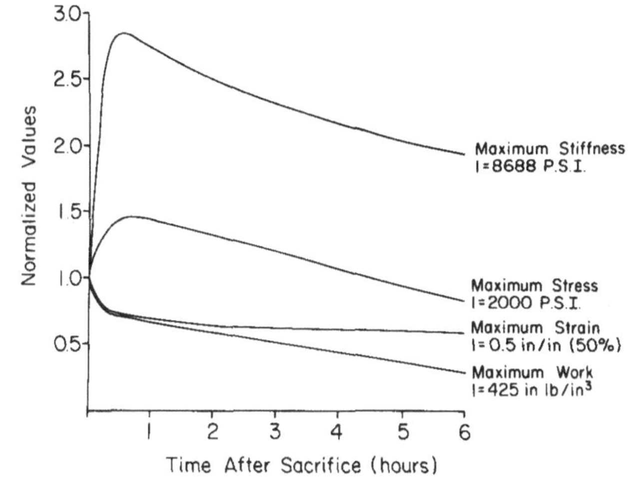

FIGURE 7. Properties of guinea pig skin as a function of time after excision at 37.8°C (normalized to immediate post-excision values). (Data extracted from Marangoni et al.[21])

The most reliable tests to date, concerning the question of the influence of post-excisional effects, are the indentation shear studies of Parsons.[23] He has determined that when viable cartilage is tested, intact on supporting subchondral bone, at 37°C in normal saline, no significant changes occur in either unrelaxed or relaxed shear moduli or spectrum of retardation times, over a period of up to 4 hr. However, storage in tissue culture media (MEM, NCTC 109 or 135) for as little as 24 hr produces significant increases in the unrelaxed modulus (of between 10 and 50%), despite no changes in apparent viability.[24-27] It is believed that this change is due to a change in the state of aggregation of the proteoglycans in the matrix, thus raising the water binding affinity.[28] This has been determined to be able to occur without significant loss of matrix.

It has also been suggested that cartilage is under an initial state of stress. This is viewed as a net tensile stress on the matrix, reflecting a dynamic balance between internal osmotic swelling pressure and the elastic quality of the collagen component of the matrix.[29] This is supposedly responsible for the familiar "split" patterns of slits that appear in articular surfaces in response to pinpoint indentations. Such forces also exist in costal cartilage. Abrahams and Duncan[30] have demonstrated the existence of a pattern of longitudinal stresses in human costal cartilage. By differential slicing, followed by quantitation of the resulting deformation (curling) of slices coupled with stress relaxation studies, they estimated that stresses as high as 200 psi exist in this tissue at excision. These are large stresses, since the average Young's modulus is on the order of 1800 psi. In addition, a considerable variability between specimens was noted.

It is widely believed that these stresses exist in vivo. The only study that might contradict this assertion is that of Parsons and Black[31] on the effect of ionicity of the bathing solution on the shear indentation behavior of viable articular cartilage on subchondral bone. They concluded that no net stress exists under conditions of bathing fluid normality. Thus, the observation of apparent stresses may be secondary to an apparent hyponormality of the environment surrounding the tissue. This can result from exposure to air (permitting evaporation) or to solutions that do not match the osmolarity and molecular weight distribution of synovial fluid.

However, if the stresses actually exist in vivo and are not artifacts of in vitro handling, then it is clear that the mere act of separating cartilage, articular or costal, from its supporting bony structures must introduce abnormal and non-physiological states of stress.

An additional problem with cartilage is the presence of large quantities of proteolytic enzymes in the chondrocytes of the tissue. Release of these enzymes, which is believed to play some role in various disease states, occurs as the tissue dies in vitro. Harris et al.[32] have showed that at 37°C, (pH 7.6 — TRIS buffer) bovine articular cartilage loses about 3%/day of hexosamine (usually taken as a measure of proteoglycan content). In their studies, this resulted in a 41% increase in strain at 15 sec under a stress of 2.45 N/cm^2 at 24 hr of incubation. This is in substantial agreement with the findings of Parsons, Simon, and Black[24] who achieved the same effect by quantitative enzymolysis, using purified hyaluronidase. Thus, autolysis can be expected to produce an increase in the spectrum of retardation times and a reduction in relaxed shear moduli in cartilage. This is clearly what happened, under the impact of papain enzymolysis, to the "floppy-eared rabbits" discussed earlier in this chapter.

It seems to be the case that cartilage will retain its properties well in vitro if temperature and ionicity are maintained, the chondrocyte population remains healthy and the tissue remains within the restraints seen in vivo. Violation of these conditions produces changes in properties, some reversible and others irreversible.

C. Bone

Bone is usually taken as a static material, with little postmortem change in behavior. However, Fitzgerald and Freeland[33] have studied the viscoelastic response of intervertebral discs as audio frequencies at both room temperature and 37°C in normal saline. They found that both the real and imaginary component of compliance fell with time after sacrifice, with an abrupt transition taking place at 5 1/2 hr after excision. Based upon this work, Fitzgerald[34] later applied the same technique to studying human bone, tooth, and rabbit bone at frequencies between 100 and 400 Hz in shear. He found the same decrease in the real compliance with time after sacrifice. His overall view of this situation was that there was a "well defined transition from living to dead cells or matter". In fact, he has called this the "life-to-death transition".

There is little literature evidence to support these studies of Fitzgerald. However, Tennyson, Ewert, and Niranjian,[35] using an ultrasonic technique (Hopkinson split bar), measured the real and imaginary moduli in alternating tension-compression as a function of postmortem age (PMA) in bovine femoral cortical bone. Although they found some reduction in the real part of the modulus between 10 and 20 days PMA, they suggested that this effect was due

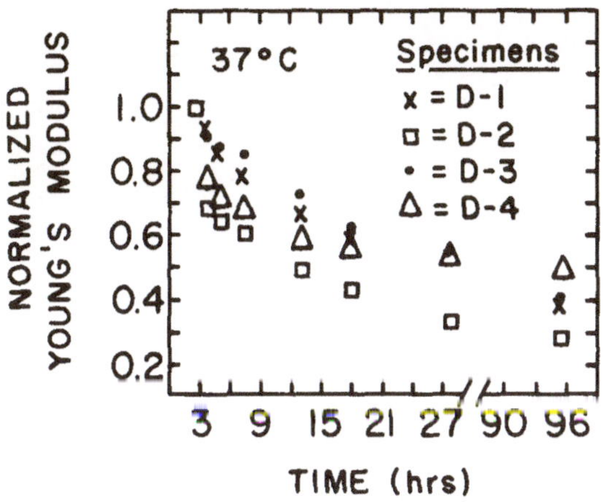

FIGURE 8. Longitudinal compressive modulus of live human cortical bone. (From Brown, N., Saputa, C., and Black, J., Young's modulus of living human bone, *Trans. ORS*, 6, 41, 1981. With permission.)

to dehydration. However, very significant changes in the loss or viscoelastic modulus were found. Back extrapolation of these changes from the value of viscosity at 3 to 10 days (3 psi·sec) suggested an in vivo value of 8 psi·sec. The indication is that the majority of this decrease takes place in the first day after excision. No mechanism was suggested by these authors.

In an attempt to verify these two studies, Brown et al.[36] studied the Young's modulus of living human bone in compression. Small blocks were taken of human tibial cortical bone, obtained at amputation. These consisted of 7 specimens from two patients. Testing in compression was initiated at from 2 to 3 hr post excision, at 37°C in tissue culture media, and the results for the storage modulus extrapolated back to zero time. The suggestion is that there is between zero and a 25% decrease in the storage modulus within the first 2 to 3 hr after excision, despite the maintenance of NVP conditions. The range of values obtained in this study was 7 to 13.5 mN/m^2. As can be seen in Figure 8, a continuous decrease was seen over the period of testing out to 96 hr. This result is clearly quantitatively in line with values previously reported for the compressive modulus of moist human bone, e.g., Evans[5] (p. 108): 8.7 Nm/m^2.

These studies suggest that changes in the viscoelastic behavior of bone do occur in the early post-excision period. However, there are not enough data available so far to permit drawing conclusions as to the direction or nature of these changes.

In the observation of fatigue in bone, the situation seems somewhat clearer. Our knowledge of the fatigue behavior of cortical bone is based almost entirely upon studies of dead bone in vitro. Carter and Hayes[37] have summarized much of what is known about the fatigue life of bone. They have developed the following general relationship:

$$\mathrm{Log}\,(2N) = -A\log(\sigma) - B\,T + C$$

where:
N = number of cycles (2N = number of reversals of load); σ = peak stress amplitude; A, B, C = constants; and T = temperature.

Data from their studies, as well as similar fatigue data from other investigators,[38-41] cited by them for bovine and human cortical bone are given in Figure 9. Here, the vertical axis is normalized stress (= maximum stress amplitude divided by 1/2 cycle ultimate tensile

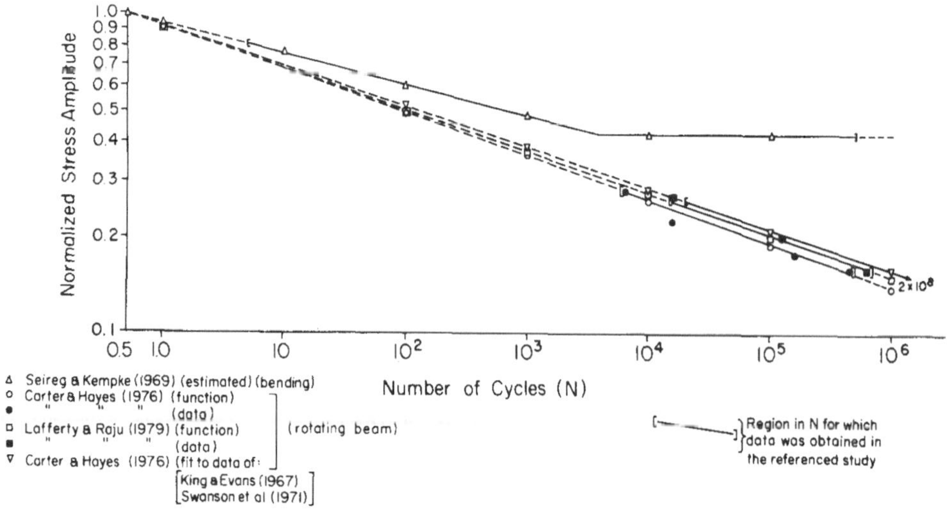

FIGURE 9. Fatigue behavior of bone in vivo and in vitro.

stress), and the horizontal axis is the logarithm of N. Experimental data are available for failure after 10^4 to 10^8 cycles. Carter et al.[42] have provided additional low cycle, high stress failure data that extend this relationship to still lower values of N (10 to 10^4).

It is clear that cortical bone, in vitro, is very sensitive to fatigue failure. Furthermore, there is no evidence of a fatigue limit in any of these studies. That is, there is no minimum stress, below which bone will not fail in fatigue, at an infinite number of cycles. The absence of such a fatigue or endurance limit is common in engineering materials that are either highly defective, e.g. filled with small defects (such as bone is) or that are tested in chemically corrosive environments.

It should be noted here that it is widely believed that a fatigue or endurance limit exists for bone in vitro. This is based upon the assertion (e.g. Kraus[43] and Evans[5]) that the work of King and Evans demonstrates an endurance limit of 4000 psi. This is actually an artifact, resulting from the erroneous conclusion that specimens tested at the lowest stress that had not failed by 10^7 cycles "would have gone on to infinity under conditions of the test" (p. 133). In fact, the predictive equation previously cited, that fits the balance of this body of data well, suggests that failure would have occurred in these specimens at about 4×10^8 cycles. Thus, there is no evidence for the existence of a fatigue limit in vitro. However, this result does not compare well with the clinical observation of fatigue endurance of man and animals in vigorous athletic activities.

Unfortunately, there exists only one study of fatigue failure in vivo. Seireg and Kempke[38] produced failure in the tibiae of rats in vivo by three-point bending. Their results, normalized by dividing load by ultimate load, are shown in Figure 9 also. Here a clear endurance limit, equal to 42% of 1/2 cycle failure load, is seen. The presence of this limit cannot be due to remodeling, since this study required only 5 1/2 hr to produce the highest number of cycles studied (6×10^5). Further evidence of the reliability is given by comparison of estimated 1/2 cycle breaking strengths of the tibiae in bending. These compare very closely with those obtained by Engesaeter et al.[44] in three-point breaking of rat tibiae at excision.

Thus it appears that live bone, in vivo, has a physical compensatory process to prevent fatigue failure at low amplitude strains in short time periods that parallels the recognized adaptive remodeling high cycle compensatory process at longer time periods. The most likely origin of this process is a protective effect of the glycosaminoglycans (GAG) ground substance of bone, which is believed to be a fluid sol in vivo as compared to a stiffer gel in vitro.[45] This gel may not be stable. Tanaka[46] et al. have shown that ionized gels can undergo abrupt collapses to as little as 1/500th of their original volumes in response to infinitesimal temperature changes or fluid phase composition. This volume contraction has the thermodynamic aspects of a true phase change. Although these observations are based upon studies of model systems, it is possible that such effects could occur in the dispersed GAG networks in mineralized tissue in response to post excision changes in internal fluid distribution and ionic concentration.

V. COMPARISON OF IN VIVO AND IN VITRO ELECTRICAL BEHAVIOR OF TISSUE

To this point I have concentrated on the mechanical aspects of tissue in examining the relationship between in vitro studies and in vivo behavior. In this section I shall briefly discuss some of the electrical aspects of bone in relation to this problem.

We recognize two types of electrical phenomena that arise in bone. The first is the passive steady dc potentials that have been called bioelectrical potentials.[47] These potentials can be measured on the surface of exposed long bone in vivo in a variety of preparations. They appear to be independent of bony landmarks or points of muscle attachment. The value or distribution in long bones is little changed by nerve denervation or nutrient artery ligation. However, local administration of blockers, or uncouplers of oxidative phosphorylation, or death of the animal produces progressive abolition of the potentials.[48]

These potentials clearly depend upon cellular viability, and removal of tissue from the body will result in progressive loss of potential magnitude and pattern as cell death is cumulative. In addition, any contribution to these potentials made by muscle injury potentials[49] produced in the preparation of the measurement site will be lost instantaneously upon excision.

A second type of potential found in bone is the piezoelectric or strain-related potential produced by tissue deformation. Although first found in vitro by many 19th century investigators, they have also been shown to occur in vivo.[50] They are found to be in phase with strain and generally to increase with driving frequency in viable bone. Voltage to strain ratios were found to be relatively strain independent, and in a NVP preparation, gradually to increase with time at constant driving power, consistent with decreasing modulus as later reported by Brown et al.[36] However, most post-excision studies have been done under much poorer conditions. Dwyer and Matthews[51] tested whole femurs from mature rats in vitro under cantilever bending. These workers compared bones fully covered with wax with unwaxed bones to examine the effects of water loss on the potential produced by a particular strain. They found a 24% decrease in peak potential in the waxed bones in 6 hr after excision, despite specific weight losses on the order of 40 mg over the test period. Studies of potential generation in situ (in cats) and also displayed similar, unexpected potential decays.

Steinberg et al.[52] pursued this observation in a series of studies on rat femora, utilizing four-point bending in vivo and very careful maintenance of hydration. They also found a significant decay of peak potential after excision, amounting to 24% in the first 6 to 7 hr and continuing to decrease, but at a slower rate thereafter. Figure 10 shows the results of a related, longer term study where intermediate periods of freezing (storage at − 15°C) were used to try to retard the phenomenon seen at 25°C. The investigators concluded that no difference was seen in the peak potential decay, although both groups showed a slight (not statistically significant) stiffening with duration of the experiment.

It is possible that these changes in the ability to produce polarization potentials in response to stress as a function of time post excision, in the absence of dehydration effects, could be related indirectly to the electrical resistivity of the bone. Neither Dwyer and Matthews nor Steinberg attempted to measure tissue resistivity. The resistivity of animal bone (rabbit)

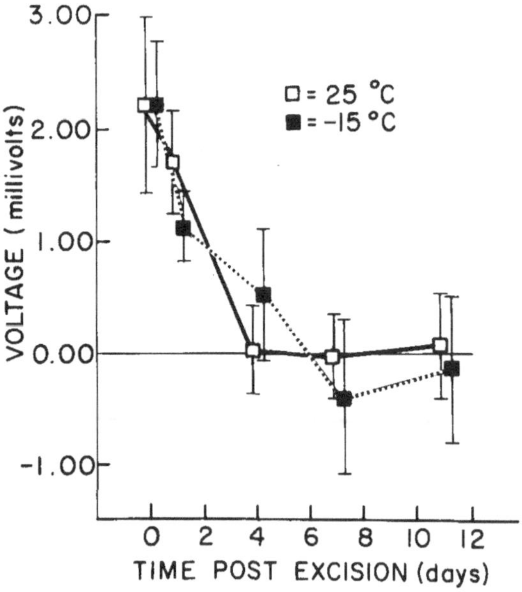

FIGURE 10. Strain-related potentials in rat femurs as a function of time and storage conditions. (From Steinberg, M. E., Finnegan, W. J., Labosky, D. A., and Black, J., Temporal and thermal effects on deformation potentials in bone. *Calcified Tiss. Res.*, 21, 135, 1977. With permission.)

in vivo is known to be on the order of 2500 ohm-cm.[53] Liboff et al.,[54] using a somewhat different preparation, found that the apparent resistivity of rabbit bone increases markedly upon sacrifice. This is not a dehydration phenomenon, as their specimens were kept fully immersed. Perhaps this increase, to resistivities on the order of 5000 to 7500 ohm-cm, is secondary to both failure of the microcirculation and to ground substance state changes (which would produce reduced ionic mobility) discussed in the previous section.

It seems clear that there is a great sensitivity of electrical behavior of bone to conditions of viability and hydration as well as time post-excision. This is an area in which much still remains to be done.

VI. CLASSIFICATION OF TISSUE CONDITIONS FOR IN VITRO STUDIES

We must recognize that, *a priori*, there are differences between living and dead tissue and that viability is an experimental parameter. I propose the use of five grades or classifications of tissue for tissue mechanical studies. These are, in decreasing order of viability:

1. VIS — Viable in situ: Tissue, with no obvious necrosis, in place in the living subject.
2. VIV — Viable in vitro: Freshly excised tissue maintained continuously in a suitable growth medium at 37°C. A majority of cells are considered to be alive.
3. NVP — Nonviable, physiologic: Tissue continuously maintained in growth media or buffered saline solution at 37°C after (transient) exposure to physical or chemical conditions that result in cell death.
4. NVM — Nonviable, moist: Tissue which has at any time been permitted to dry out or which has been cooled below 25°C and then rehydrated or reheated to 37°C.

Further departure from in vivo conditions, such as preparation of calcified tissue for study by drying and defatting, creates conditions that cannot be related, at present, to those in vivo. Such preparations are required for certain experiments. The tissue classification would be:

5. NVD — Nonviable, dry: dehydrated or defatted tissue.

These grades provide a framework for the interpretation of the departure of the tissue from its true in vivo state. It is important to remember that while tissue may definitely be considered dead, that is, with an absence of any cellular function, it cannot be similarly considered absolutely alive.

The relative viability of tissue of the VIV type will be affected by many factors. The failure of the microcapillary system and a resultant failure of infusion from the surrounding medium will result in progressive cell death. Thus, it is important to ascertain the time after excision of tissue in any of these classes. The use of one of the various methods of measuring metabolic events such as O_2 uptake rate, $SO_4^{=}$ incorporation, ATP, or other obligate material production rate, and comparison of these rates to those found in vivo (VIS) will provide both a quantitative comparison and information on the time course of viability.

It should have been firmly established by now that the post-excision time affects mechanical test results. It has not been, because the most important changes occur in the first 6 to 12 hr postexcision for NVP or NVM tissue, the excised material types most commonly studied. Most mechanical studies do not begin this quickly after excision due to the problems of sample preparation, equipment stabilization, etc. If they do begin, early data is often discarded as unreliable, that is, nonreproducible. The most convenient data, that which can be reproduced, is obtained from material that has been excised 24 to 72 hr prior to experimentation.

While it is probably desirable to attain conditions as close to VIS as possible, that is, VIV, this is often not logistically possible. The alternative, particularly for biological experiments where we wish to distinguish between treatments, rather than to obtain absolute results, is to attempt to normalize post-excisional effects. This can be done by adopting an experimental design based upon isochronal testing. That is, each specimen is treated identically and tested at the same time post excision. In the case of the tendon excursion studies discussed previously,[18] this was achieved by preparing the mechanical preparation in an anesthetized animal, randomizing the order of preparation of experimental and control sides, and then sacrificing the animal for testing after the second preparation was removed. Similar strategies can be built into other types of studies. Furthermore, despite the problems associated with frozen storage, there is sound evidence in several studies (for instance, Marangoni et al.[21]) that rapid, isochronal freezing followed by controlled thawing can produce well-controlled results, with, in many cases, good correspondence to immediate post-sacrifice data.

What is unsatisfactory is the method implied by the following description: " . . . After sacrifice, the tendons were excised and stored in saline for periods of 1/2 to 7 days before testing, except for Group C, which was frozen. We cycled each specimen until reproducible data was obtained; Table 4 gives the mean of several tests for each tendon."

Except for the fact that this "quotation" was constructed to illustrate a point, it might have come, in nearly this form, from any one of a large number of papers on tissue mechanics.

VII. CONCLUSIONS

It would be inviting to recapitulate the evidence of post-excisional changes in tissue properties cited here, as well as the vast literature that has been overlooked, both inadvertently or deliberately. To do so would be to assume a degree of certainty concerning the general theses advanced here that is unwarranted. Thus, conclusions will be restricted to two general injunctions:

1. Tissues display native properties only when viable, in situ, and when deformed, qualitatively and quantitatively, within physiological ranges.
2. Deviation from these conditions produces a variety of changes in properties, some reversible and others irreversible, that can be understood only by careful documentation of changes in physical conditions and of the passage of time from viable in situ conditions.

If those interested in experimental procedures with tissue would take these two injunctions to heart, we can hope in the future to be able progressively to sort out the meaning of such inexact concepts as "pre-conditioning", "biological variability", etc. Failure to do so will perpetuate the process of retrospective falsification that has dominated much of tissue mechanical investigation to date.

REFERENCES

1. **Barber, B. and Fox, R. C.,** The case of the floppy-eared rabbits: an instance of serendipity gained and serendipity lost, *Am. J. Soc.*, 64, 128, 1958.
2. **Thomas, L.,** Reversible collapse of rabbit ears after intravenous papain and prevention of recovery by cortisone, *J. Exp. Med.*, 104, 245, 1956.
3. **Kellner, A., Robertson, T., and Mott, H. O.,** Blood coagulation defect induced in rabbits by papain solutions injected intravenously, *Fed. Proc.* X, 1, 1951.
4. **Sedlin, E. D. and Hirsch, C.,** Factors affecting the determination of the physical properties of femoral cortical bone, *Acta Orthop. Scand.*, 37, 29, 1966.
5. **Evans, F. G.,** *Mechanical Properties of Bone*, Charles C Thomas, Springfield, Ill., 1973.
6. **Stevens, J. and Ray, R. D.,** An experimental comparison of living and dead bone in rats. I. Physical Properties, *J. Bone J. Surg.*, 44B, 412, 1962.
7. **Lissner, H. R. and Roberts, V. L.,** Evaluation of skeletal impacts of human cadavers, in *Studies on the Anatomy and Function of Bone and Joints*, Evans, F. G., Ed., Springer-Verlag, Heidelberg, 1966, 113.
8. **Greenberg, S. W., Gonzalez, D., Gurdjian, E. S., and Thomas, L. M.,** Changes in physical properties of bone between the *in vivo*, freshly dead and embalmed conditions, *Proc. 12th Stapp Car Crash Conf.*, Society of Automotive Engineers, New York, 1968, 271.
9. **Gurdjian, E. S. and Lissner, H. R.,** Deformation of the skull in head injury. A study with the "stress coat" technique, *Surg. Gynecol. Obstet.*, 1, 679, 1945.
10. **McElhaney, J., Fogle, J., Byars, E., and Weaver, G.,** Effect of embalming on the mechanical properties of beef bone, *J. Appl. Physiol.*, 19, 1234, 1966.
11. **Wall, J. C., Chatterji, S., and Jeffery, J. W.,** On the original scatter in results of human bone strength tests, *Med. Biol. Eng.*, 8, 171, 1970.
12. **Viidik, A., Sandquist, L., and Magi, M.,** Influence of postmortal storage on tensile strength characteristics and histology of rabbit ligaments, *Acta. Orthop. Scand.*, (Suppl. 79), 1965.
13. **Viidik, A. and Lewin, T.,** Changes in tensile strength characteristics and histology of rabbit ligaments induced by different modes of postmortal storage, *Acta. Orthop. Scand.*, 37, 141, 1966.
14. **Tkaczuk, H.,** Tensile properties of human lumbar longitudinal ligaments, *Acta. Orthop. Scand.* (Suppl.), 115, 1968.
15. **Matthews, L. S. and Ellis, D.,** Viscoelastic properties of cat tendon: effects of time after death and preservation by freezing, *J. Biomech.*, 1, 65, 1968.
16. **Benedict, J. V., Walker, L. B., and Harris, E. H.,** Stress-strain characteristics and tensile strength of unembalmed human tendon, *J. Biomech.*, 1, 53, 1968.
17. **Chu, B. M., Frasher, W. G., and Wayland, H.,** Hysteretic behavior of soft live animal tissue, *Ann. Biomed. Eng.*, 1, 182, 1972.
18. **Black, J.,** Dead or alive: the problem of *in vitro* tissue mechanics, *J. Biomed. Mater. Res.*, 10, 377, 1976.
19. **Barbanel, J. C., Evans, J. H., and Finlay, J. B.,** Stress-strain time relations for soft connective tissues, in *Perspectives in Biomedical Engineering*, Kenedi, R. M., Ed., University Park Press, Baltimore, 1973, 165.
20. **Finlay, B.,** Dynamic mechanical testing of human skin "in vivo", *J. Biomech.*, 3, 557, 1970.

21. **Marangoni, R. D., Glaser, A. A., Must, J. S., Brody, G. S., Beckwith, T. G., Walker, G. R., and White, W. L.,** Effect of Storage and handling techniques on skin tissue properties, *Ann. N.Y. Acad. Sci.*, 136, 439, 1966.
22. **Smith, J. W.,** The elastic properties of the anterior cruxiate ligament of the rabbit, *J. Anat. (Brit.)*, 88, 369, 1954.
23. **Parsons, J. R. and Black, J.,** The viscoelastic shear behavior of normal rabbit articular cartilage, *J. Biomechanics*, 10, 21, 1977.
24. **Parsons, J. R., Simon, W. H., and Black, J.,** Viscoelastic shear response of viable articular cartilage after quantitative enzymolysis: hyaluronidase, *Trans. ORS*, 2, 78, 1977.
25. **Steinberg, M. E., Labosky, D. A., Parsons, J. R., and Black, J.,** Changes in mechanical properties of articular cartilage in antigen induced arthritis, *Trans. ORS*, 2, 79, 1977.
26. **Steinberg, M. E., Parsons, J. R., Koh, J. K., Simon, W. H., Black, J., and Lane, J.,** Articular cartilage changes caused by D-Penicillamine, *Trans. ORS*, 1, 97, 1976.
27. **Black, J., Shadle, C. A., Parsons, J. R., and Brighton, C. T.,** Articular cartilage preservation and storage: II. Mechanical indentation testing of viable stored articular cartilage, *Arthritis Rheum.*, 22, 1102, 1979.
28. **Parsons, J. R. and Black, J.,** On the thermodynamics of the viscous deformational mechanism of articular cartilage, *Trans. SFB*, 1, 78, 1977.
29. **Maroudas, A.,** Balance between swelling pressure and collagen tension in normal and degenerate articular cartilage, *Nature*, 260, 808, 1976.
30. **Abrahams, S. M. and Duncan, T. C.,** The mechanical characteristics of costal cartilage, in *Biomechanics and Related Bio-Engineering Topics*, Kenedi, R. M., Ed., 285, 1965.
31. **Parsons, J. R. and Black, J.,** Mechanical behavior of articular cartilage: quantitative changes with alteration of ionic environment, *J. Biomech.*, 12, 765, 1979.
32. **Harris, E. D., Jr., Parker, H. G., Radin, E. L., and Krane, S. M.,** Effects of proteolytic enzymes on structural and mechanical properties of cartilage, *Arthritis Rheum.*, 15, 497, 1972.
33. **Fitzgerald, E. R. and Freeland, A. E.,** Viscoelastic response of intervertebral disks at audiofrequencies, *Med. Biol. Eng.*, 9, 459, 1970.
34. **Fitzgerald, E. R.,** Postmortem transition in the dynamic mechanical properties of bone, *Med. Phys.*, 4, 49, 1977.
35. **Tennyson, R. C., Ewert, R., and Niranjian, V.,** Dynamic viscoelastic response of bone, *Exper. Mech.*, 12, 502, 1972.
36. **Brown, N., Saputa, C., and Black, J.,** Young's modulus of living human bone, *Trans. ORS*, 6, 97, 1981.
37. **Carter, D. R. and Hayes, W. C.,** Fatigue life of compact bone — I. Effects of stress amplitude, temperature and density, *J. Biomechanics*, 9, 27, 1976.
38. **Seireg, A. and Kempke, W.,** Behavior of *in vivo* bone under cyclic loading, *J. Biomechanics*, 2, 455, 1969.
39. **Lafferty, J. F. and Raju, P. V. V.,** The influence of stress frequency on the fatigue strength of cortical bone, *J. Biomed. Eng.*, 101, 112, 1979.
40. **Swanson, S. A. V., Freeman, M. A. R., and Day, W. H.,** The fatigue properties of human cortical bone, *Med. Biol. Eng.*, 9, 23, 1971.
41. **King, A. I. and Evans, F. G.,** Analysis of fatigue strength of human compact bone by the Weibull technique, in *Digest of the Seventh International Conference on Medical and Biological Engineering*, Jacobson, B., Editor, The Organizing Committee of the Conference, Stockholm, 1967, 514.
42. **Carter, D. R., Caler, W. E., Spengler, D. M., and Frankel, V. H.,** Uniaxial fatigue of human cortical bone: the influence of tissue physical characteristics, *J. Biomech.*, 14, 461, 1981.
43. **Kraus, H.,** On the mechanical properties and behavior of human compact bone, *Adv. in Biomed. Eng.*, 2, 169, 1968.
44. **Engesaeter, L. B., Ekeland, A., and Langeland, N.,** Methods for testing the mechanical properties of the rat femur, *Acta. Orthop. Scand.*, 49, 512, 1978.
45. **Gersh, I. and Catchpole, H.,** Nature of ground substance in connective tissue, *Perspect. in Biol. Med.*, 3, 282, 1959.
46. **Tanaka, T., Fillmore, D., Sun, S. T., Nishio, I., Swizlow, G., and Shah, A.,** *Phys. Rev. Letts.*, 45, 1636, 1980.
47. **Friedenberg, Z. B. and Brighton, C. T.,** Bioelectric Potentials in bone, *J. Bone J. Surg.*, 48A, 915, 1966.
48. **Friedenberg, Z. B., Harlow, M. C., Heppenstall, R. B., and Brighton, C. T.,** The cellular origin of bioelectric potential in bone, *Calc. Tiss. Res.*, 13, 53, 1973.
49. **Lokietek, W., Pawluk, R. J., and Bassett, C. A. L.,** Muscle injury potentials: a source of voltage in the undeformed rabbit tibia, *J. Bone J. Surg.*, 56B, 361, 1974.

50. **Black, J.,** Strain related potentials in viable human cortical bone, Thesis, University of Pennsylvania, 1972.
51. **Dwyer, N., S. J. P. and Matthews, B.,** The electrical response to stress in dried, recently excised and living bone, *Injury*, 1, 279, 1970.
52. **Steinberg, M. E., Finnegan, W. J., Labosky, D. A., and Black, J.,** Temporal and thermal effects on deformation potentials in bone, *Calcified Tiss. Res.*, 21, 135, 1977.
53. **Black, J., Marcum, S., and Brighton, C. T.,** The low frequency electrical resistivity of rabbit tissues *in vivo*, *Trans. ORS*, 5, 230, 1980.
54. **Liboff, A. R., Rinaldi, R. A., Lavine, L. S., and Shamos, M. H.,** On electrical conduction in living bone, *Clin. Orthop.*, 106, 330, 1975.

Chapter 3

STRUCTURE AND FORMATION OF BONE MINERAL

Adele L. Boskey and Aaron S. Posner

TABLE OF CONTENTS

I. INTRODUCTION

Bone is composed of mineral and mineral ions, water and an organic matrix. Although the relative proportions of mineral, water, and matrix molecules of bone vary among types, species, and development ages of bone, the overall chemical composition of bone is amazingly constant (Table 1). Most of the macromolecules and mineral in bone occur extracellularly. It is the interaction of the cells with the extracellular matrix, as well as the interactions of all of the extracellular matrix components, which enable bone to perform its unique functions, providing mechanical strength and maintaining calcium and phosphorus homeostasis.

Each of the chemical entities found in bone has one or more specific functions. Many of the macromolecules found in bone are involved in controlling the deposition of mineral, while others are involved in mineral homeostasis or supplying nutrition to the tissue. We will review the nature of the bone components and how they are peculiarly suited to bone function. An examination of the anatomical structure of bone and a comparison of the distribution of macromolecules in each anatomical region during growth provides a basis for understanding the functions of these macromolecules. A knowledge of the microscopic anatomy of bone development is also useful in understanding the chemistry of this process. We will, therefore, first define the morphological changes which occur as bone forms, then discuss the nature of bone mineral, and finally, review the chemistry and function of the organic matrix of bone.

All of the bones in the body function both to provide mechanical strength and protection, and to regulate mineral homeostasis. Thus, although morphologically a bone appears to consist of different tissue types, all of these tissues act together as an organ. Morphologically, there are two major types of bony tissue, cancellous (spongy) bone and cortical (compact) bone (Figure 1). Cortical bone, found in the shafts of long bones, is denser and more highly ordered, macroscopically and microscopically, than the randomly-woven cancellous bone.

To appreciate the functions of the macromolecules found in these bones, it is important to understand the two processes by which organs develop: intramembranous growth and endochondral ossification.

In the endochondral process, bone forms on a framework of calcified cartilage. In the case of intramembranous bone formation, bone is deposited directly by cells which secrete a matrix (osteoid) which mineralizes in situ without going through a cartilaginous stage. Thus, the long bones (weight-bearing bones) are strengthened during endochondral growth by the presence of a layer of calcified cartilage, which forms an interlocking mesh between the bone and the non-calcified tissue. In contrast, membranous bone, such as the bones of the skull, forms by apposition of new mineralized layers on the organ surface, first forming spongy interior layers, which in time, are covered by harder, cortical layers. It is important to note that the major difference between these two processes, endochondral ossification and intramembranous bone formation, is in the environment in which these processes occur and not in the kind of bone formed.[1]

During bone development the cells secrete and regulate the deposition and turnover of an extracellular matrix upon which mineral will be deposited. The changes that occur both in the extracellular matrix and in the cells prior to mineral deposition create circumstances favorable for mineralization. For example, the epiphyseal growth plates at the ends of growing long bones are regions of active endochondral ossification (Figure 2).[2] Here, at the extreme ends of the growth plate, the small round reserve or resting cartilage cells (chondrocytes) proliferate, forming rapidly dividing columnar (proliferating/palisading) cells. These proliferating cells are the most metabolically active of the growth plate cartilage cells, having an extremely rapid rate of DNA synthesis. The major protein synthesized by these chondrocytes is cartilage (type II) collagen, but these cells also synthesize proteoglycans and certain cartilage enzymes (see below). After the cells proliferate they begin to swell, quadrupling

Table 1
THE COMPOSITION OF VARIOUS BONES CALCULATED FROM PUBLISHED DATA

Source	Age	Type	%H_2O	Percent of dry wt: Mineral	Collagen	GAG[†]	Lipid
Cow	calf	periosteal	*	54.0[a]	22.2[a]	0.38[a]	*
		cancellous	*	52.0[a]	19.4[a]	0.52[a]	1.7[b]
		cortical	*	63.0[b]	16.9[b]	*	0.6[b]
	adult	cortical	9.1[c]	76.4[c]	21.5[c]	0.20[c]	1.03[c]
Human	fetal	cortical	*	70.8[d]	18.3[d]	0.53[d]	*
	2—13 yr	cortical	*	73.7[d]	17.7[d]	0.41[d]	*
	adult	cortical	7.3[c]	67.2[c]	21.2[c]	0.34[d]	1.6[f]
		membranous	15.2[e]	67.4[e]	21.1[e]	0.28[e]	*
Rat	weanling	cortical	*	68.4[f]	19.0[f]	0.5[f]	1.2[f]
		cancellous	*	54.5[f]	21.0[f]	0.4[f]	2.8[f]
	adult	cortical	12.0[f]	69.0[g]	20.0[g]	0.61[f]	2.5[g]
		cancellous	*	54.0[g]	15.7[g]	0.69[f]	3.2[g]

† Glycosaminoglycan.
* Not determined.

[a] Reference 57.
[b] Reference 52.
[c] Reference 58.
[d] Reference 59.
[e] Reference 60.
[f] Unpublished data.
[g] Reference 53.

in volume. It is in the areas adjacent to the last of these hypertrophic chondrocytes that calcification of the extracellular cartilage matrix takes place. Then, there is an invasion of blood vessels into the calcified cartilage. This results in the removal of the initial calcified deposits, and the concomitant deposition of woven bone (primary spongiosa) by osteoblasts (bone forming cells) on a type I (bone) collagen matrix. The importance of vascular invasion for triggering this osteogenesis has long been known.[3] The earliest formed woven bone will subsequently be remodeled to yield cortical bone with a well-developed marrow cavity.

Certain additional morphological changes can be seen in the cartilage matrix during the process of endochondral ossification. In the proliferating zone the cells are particularly rich in rough endoplasmic reticulum, indicative of their high metabolic activity. The mitochondria of the chondrocytes in the proliferating zone contain numerous electron dense granules while the cells in the more distal zone of provisional calcification contain few such granules.[4] X-ray diffraction analysis indicated these mitochondrial granules are composed of amorphous tricalcium phosphate.[5] The disappearance of calcium phosphate from the mitochondria at the time that mineral crystals first appear in the extracellular matrix suggests a causal relationship between these intra- and extracellular deposits.

Much of the first mineral to appear outside the chondrocytes is associated with membrane-bound bodies called extracellular matrix vesicles. These vesicles are believed to provide a preferential site for mineral deposition in the growth plate, as well as in other calcified tissues.[6-8]

There are other types of bone development that also occur by endochondral ossification. The chemical and morphological events which take place in the healing of a fractured bone mimic the events of endochondral ossification. The cartilaginous callus which forms first, is calcified, and then is replaced by woven bone.[9] Bone induction by the subcutaneous

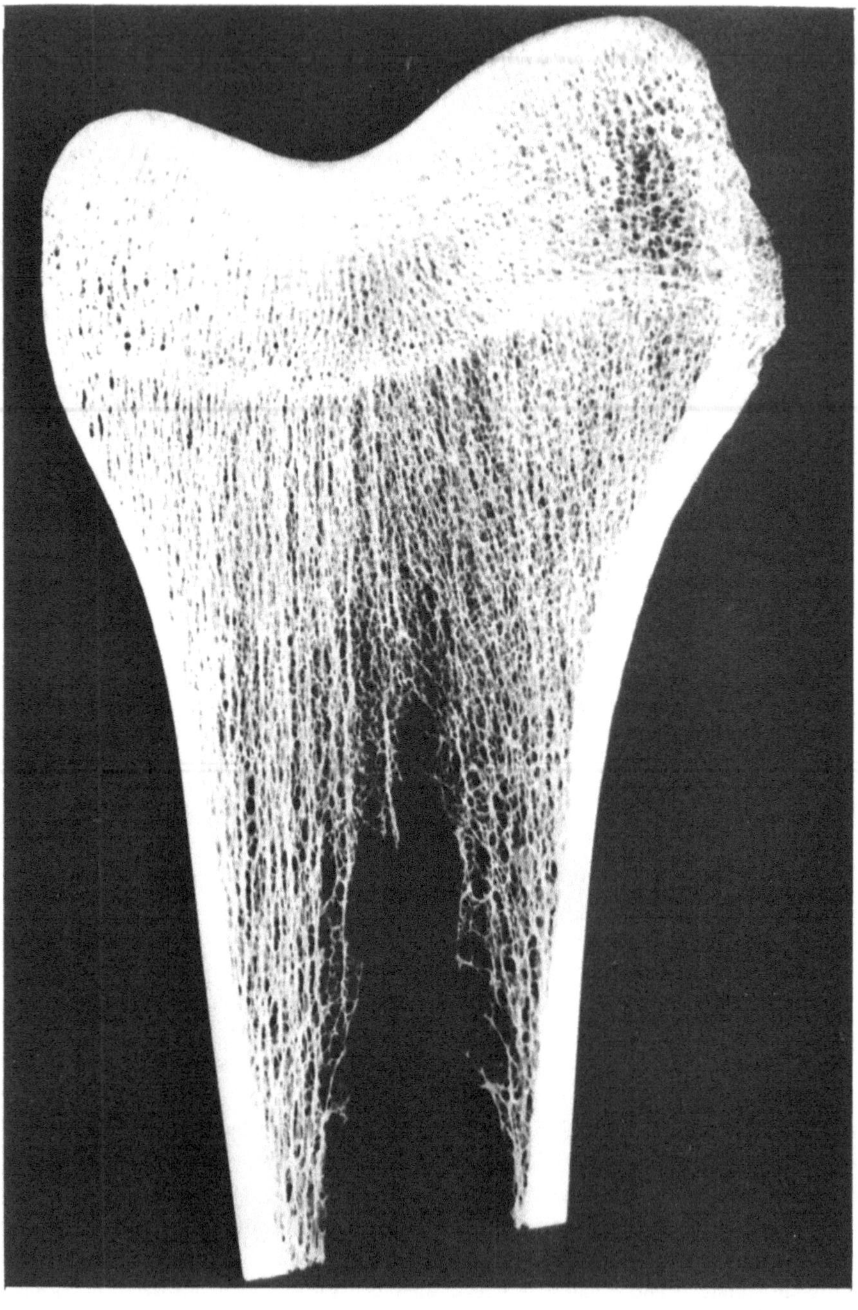

FIGURE 1. Radiograph of a longitudinal section of the head of an adult human femur showing the dense (radio-opaque) cortical bone surrounding the marrow cavity and the mesh-like cancellous bone with fine trabeculae. (Courtesy of Dr. Peter G. Bullough).

implantation of demineralized bone also proceeds through a process of endochondral ossification.[10] In bone induction the morphological and biochemical changes take place homogenously throughout the matrix as a function of time. Table 2 lists the temporal changes during bone induction as an illustration of the stages of endochondral ossification.

Mineral deposition in endochondral ossification occurs first on a cartilage matrix. This matrix and the earliest crystals are then replaced by a bone matrix and bone mineral. In the

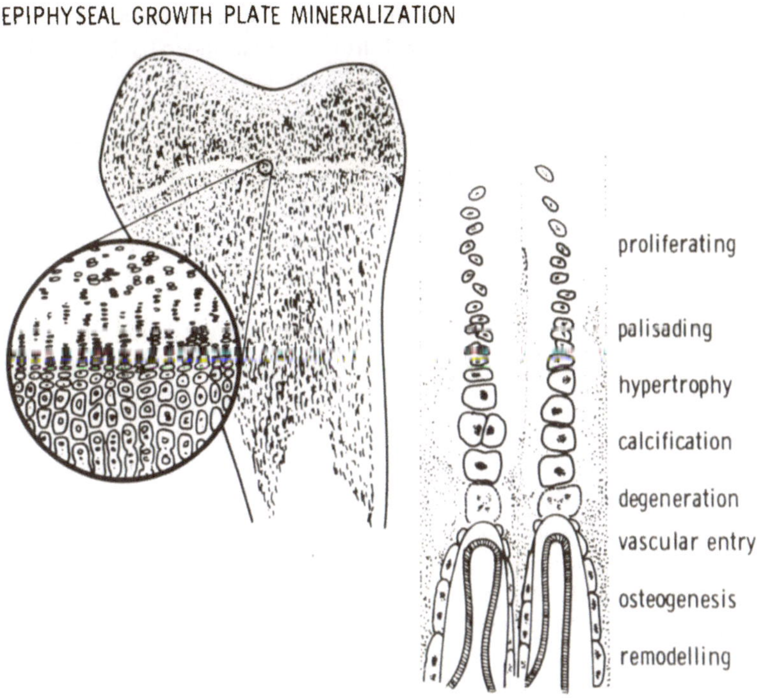

FIGURE 2. Diagram of epiphyseal growth plate (white horizontal strip) of a long bone. Microscopic views of this region, which are shown in two increasing magnifications, illustrate changes in the mineralization process.

Table 2
BIOSYNTHETIC CAPABILITIES OF EVOLVING CELLS DURING BONE-MATRIX INDUCED ENDOCHONDRAL OSSIFICATION[a]

Days after implantation	Cell types present	Macromolecules synthesized
1	Polymorphonuclear leukocytes	—
3	Mesenchymal cells	Type III collagen; fibronectin
5	Chondroblasts	Types I and III collagen; low mol wt proteoglycans
7	Chondrocytes	Types II and III collagen; high mol wt proteoglycans
9	Hypertrophic chondrocytes; multinucleated chondroclasts	Types I, II, and IV collagen; decreased proteoglycan production
10—11	Osteoclasts; osteoblasts; macrophage hematopoetic stem cells	Types I, II, III, and IV collagen; proteoglycan; bone-GLA protein; hemoglobin

[a] Adapted from Reference 10.

membranous bones, even the earliest mineral forms on a bone matrix. In the following sections we shall see that the nature of the mineral in each of these tissues is quite similar. We will then examine the differences between cartilage and bone matrices, and review how mineral deposition occurs in these different matrices.

II. BONE MINERAL

Mature (post-embryonic) bone mineral is an analogue of the calcium phosphate compound hydroxyapatite. Bone apatite is characterized by a high specific surface, carbonate substitution, non-stoichiometry and internal crystal disorder. Our understanding of the formation and nature of bone mineral has come not only from studies on the tissue itself but from work on both well-crystallized and finely-divided hydroxyapatites. Chemically and biologically precipitated hydroxyapatites are so small in crystal size that it has been necessary to use x-ray and neutron diffraction studies on well-crystallized analogues to delineate certain important features of the finely-divided biological apatites. Such diffraction studies[11] gave us the structural parameters of the constituent atoms of the hydroxyapatite unit cell, $Ca_{10}(PO_4)_6(OH)_2$, which were needed to interpret later studies on the poorly crystallized hydroxyapatites.

Studies on large crystals have led to a better understanding of the role of carbonate as well as other isomorphous substitutions found in bone mineral. Infrared and x-ray studies suggested that the planar CO_3 ion substitutes, in some undefined way, for the tetrahedaral PO_4 group in hydroxyapatite prepared at or below body temperature.[12] Infrared and x-ray diffraction studies have shown that carbonate, substituting for PO_4 groups, induces atomic misalignment in both poorly crystallized and well-crystallized apatites.[13,14] This structural distortion is an important feature of bone apatite contributing to its high chemical reactivity.

In vitro studies have provided insight into how bone mineral forms. When hydroxyapatite is precipitated from a highly supersaturated calcium phosphate solution an unstable x-ray amorphous calcium phosphate precursor phase is observed. This non-crystalline material has a Ca/P molar ratio of 3:2, rather than the 5:3 of hydroxyapatite. It converts autocatalytically in solution to a finely-divided, non-stoichiometric hydroxyapatite which ripens slowly improving in stoichiometry and crystal size.[15] When hydroxyapatite is precipitated from solutions of low supersaturation, very fine dot-like crystals of hydroxyapatite, smaller than those formed at high supersaturation, precipitate directly, with no amorphous precursor.[16] However, there is evidence that the earliest extracellular mineral deposited in chick embryo is a non-crystalline calcium phosphate.[17] Some workers report that the first phase which appears in the mineralization of extracellular matrix vesicles in the growth plates of long bones is an x-ray amorphous calcium phosphate,[18] while others report hydroxyapatite as the first phase.[19,20] Others[21] have suggested that in early bone formation in the chick embryo the mineral brushite, $CaHPO_4{\cdot}2H_2O$, is a precursor to bone apatite. More work remains to be done to determine the nature of the first mineral phase depositing in bone formation. The identification of the earliest mineral phase will define the chemical milieu (high supersaturation, low supersaturation, acid pH, etc.) in which the mineral forms. However, it should be noted that most of the mineral deposited in ossification forms by growth on existing crystals. Thus, although we may not be certain of the nature of the first mineral, studies of hydroxyapatites are highly suitable for characterizing bone mineralization.

Bone apatite and other precipitated hydroxyapatites are distinct from high temperature apatite preparations because of their submicroscopic crystal size, structural distortion, and low Ca/P ratios. Figure 3 shows a comparison of x-ray diffraction (left) and infrared absorption (right) data from a standard well crystallized hydroxyapatite and mature human bone. The small crystal size and crystal distortion in bone apatite (A) results in the poorly resolved x-ray pattern as compared to the synthetic standard pattern (B). The details of the infrared spectra show that the OH stretching (3572 cm^{-1}) and OH librational (630 cm^{-1}) modes are missing in the bone pattern (A) but present in standard pattern (B), although OH is present in both materials. In addition, the splitting of the P-O bending mode doublet at 550 to 600 cm^{-1} is less well resolved in bone than in the standard. Both of these infrared effects are attributable to the small size and/or distortion in bone apatite.

A number of studies have concluded that the low Ca/P ratios observed in precipitated hydroxyapatites results from structurally missing calcium ions. A general formula was suggested for non-stoichiometric hydroxyapatites: $Ca_{10-x}H_x(PO_4)_6(OH)_{2-x}$, with x varying from

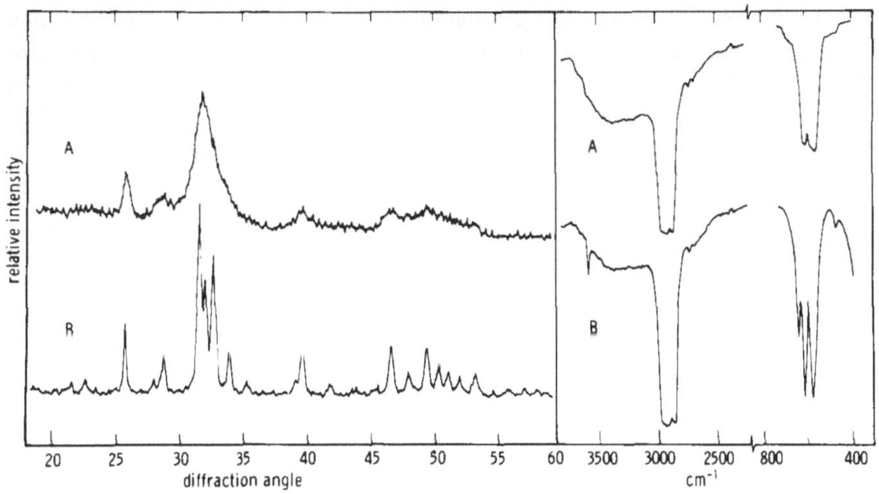

FIGURE 3. Comparison of x-ray diffraction (left) and infrared adsorption (right) patterns from mature bone (A) and well-crystallized hydroxyapatite (B). Copper K α radiation in x-ray experiment; diffraction angle in degrees 2θ. The intense infrared bands at 2900 cm^{-1} are due to Nujol® used in sample preparation.

zero (stoichiometric) to almost 2. This formula was based on the amount of pyrophosphate produced when a series of Ca-deficient hydroxyapatites of differing Ca/P were heated to 600°C.[22] Recent titration experiments[23] showed directly that precipitated Ca-deficient hydroxyapatites contain such OH deficiencies.

The production of pyrophosphate above 600°C occurs only in Ca-deficient apatites and not in perfect apatites. When Ca-deficient hydroxyapatites are heated above 800°C, (a temperature which also does not affect stoichiometric hydroxyapatite), they dehydrate to form β-$Ca_3(PO_4)_2$ plus some residual apatite. If the Ca/P ratio is equal to or less than 1.5, there is no residual apatite seen after pyrolysis at 800°C. Bone apatite produces pyrophosphate when heated at low temperature and β-$Ca_3(PO_4)_2$ when heated at higher temperature, in the amounts which suggest that bone mineral is about 10% deficient in calcium.[24]

The submicroscopic bone mineral crystals are constantly being resorbed and reformed at a high rate during body growth and at a much slower rate after maturity. A number of factors exert a significant effect on the solubility of bone apatites allowing their removal. The Gibbs-Kelvin equation predicts that the solubility of crystals increases as their specific surface increases (i.e., their size decreases). In the size range observed for bone apatite, the solubility can vary considerably due to this effect; a small increase in size can effect a measurable reduction in solubility.

The size and shape of bone apatite crystals have been measured by direct observation using the electron microscope and by indirect calculations employing x-ray diffraction methods. The literature on this subject is confusing, for the size and shape of the mineral particles change with species, age, and disease state; in addition, a single specimen contains a range of particle sizes and shapes. A representative study[25] of bone from femurs of human, ox and rabbit, suing both x-ray diffraction and dark field electron microscopy, concluded that bone mineral particles are probably plate shaped. These bone crystallites measured 50 to 60 Å in their smallest dimension, and the largest dimension had a mean value of 320-360, Å, although some crystals were 1000 Å in length.

The high reactivity of bone mineral is due in part to crystal imperfection as well as to crystal size. As noted, above, CO_3-substitution and Ca-deficiencies distort apatite structures, resulting in crystal strain, which in turn raises their apparent solubility. The adsorption of various molecules on the surface of the apatite crystals can also affect the solubility of the mineral. Pyrophosphate ions and certain diphosphonates, for example, have been shown[26] to inhibit both the nucleation of newly forming apatites as well as the dissolution of already formed apatites. These inhibitors act by adsorbing on active growth or dissolution sites on the newly formed, or already formed crystal, blocking access to the solution. In addition, various macromolecules can bind to the surface of apatite crystals decreasing the available apatite surface area and/or active sites. It is apparent that the dissolution of small crystals of hydroxyapatite is a complex phenomenon that is affected by many variables not yet completely understood.

The formation of mineral in the extracellular matrices of cartilage and bone involves the deposition of initial mineral crystals, followed by regulated growth and further proliferation of the mineral. Three general processes occur in calcifying tissues which do not occur in non-calcifying tissues. First, calcification inhibitors, present in all connective tissue, can be removed or modified in order for hydroxyapatite deposition to occur. Second, the local calcium and phosphate activities can be increased at sites of mineral deposition to support nucleation and growth. Third, heterogeneous nucleators which induce hydroxyapatite formation may be formed or activated. These events may occur individually or in combination. After formation of the first hydroxyapatite crystals at multiple sites in the extracellular bone matrix, we observe subsequent crystal deposition, growth and orientation which are regulated, in part, by specific organic moieties. The nature and role of the organic matrix of bone will be discussed in the section that follows.

III. THE ORGANIC MATRIX

A. Collagen

The protein collagen[27] accounts for 70 to 90% of the non-mineralized components of bone. A major protein of the body, collagen occurs in a variety of tissues, both calcifying and non-calcifying. The collagen molecule consists of three polypeptide (alpha) chains coiled about one another in a triple helix. Individual collagen molecules, about 300 nm in length, are assembled in highly ordered fibrils which are stabilized by inter- and intra-molecular crosslinks. The triplet amino acid sequence, glycine-X-Y, occurs repeatedly throughout each alpha chain; X is frequently proline, Y is often hydroxyproline or hydroxylysine. The presence of the smallest amino acid, glycine, at every third residue is essential for forming the triple helix. Proline and hydroxyproline are present for the maintenance of structural stability. Hydroxylysine and glycosylated hydroxylysine are involved in the crosslinks which stabilize the polymeric, collagen fibrils.

Although they all contain the primary $(GLY\text{-}X\text{-}Y)_n$ structure, the collagens found in different tissues are composed of dissimilar alpha chains. To date, at least seven unique alpha chains contained in five types of collagen have been identified as distinct gene products.[27] The type I collagen of bone, skin, dentin and tendon contains two identical alpha and one dissimilar alpha chains ($(\alpha(I)_2\alpha2)$). The type II collagen of cartilage contains three identical chains ($(\alpha(II)_3)$) which are different in amino acid sequence from those in type I collagen.

Bone and dentin type I collagens are the major proteins of mineralized tissues, but type I collagen is also found in tissues that do not normally mineralize such as skin. Thus, many investigators have looked for structural differences among the type I collagens which could explain their presence in tissues of differing properties. The major differences among type I collagens are related to the distribution and nature of the crosslinks which are responsible for the stability and arrangement of the collagen fibrils. It should be stressed that not only type I collagen mineralizes, for, as was seen above, type II collagen mineralizes in endochondral calcification. In this light, indirect immunofluorescence studies[28,29] have identified both type I and type II collagens as sites of mineral deposition in endochondral ossification. Furthermore, during matrix-induced bone formation, both types I and II collagen have been detected by chemical analyses at the time of early bone mineralization.[30]

In general, the collagen fibrils provide the framework upon which the mineral is deposited in bone. The c-axis of bone hydroxyapatite crystals is always parallel to the fiber axis of the collagen.[31] The distribution of the crosslinks in bone collagen as well as the weave of the bone collagen fibrils are said to influence the calcifiability of bone collagen regulating the extent of mineralization and the crystal size of the mineral.[32]

Early studies, reviewed elsewhere,[32] suggested that bone collagen acted as a heterogeneous nucleator of hydroxyapatite. It is currently believed that collagen itself is not an apatite nucleator, but other substances closely associated with collagen serve this function. However, the presence of collagen is essential for bone formation and development. In fact, a mineralized tissue without a collagenous matrix is not bone. Related to this, is the observation that errors in the biosynthesis of type I collagen lead to skeletal abnormalities.[33] This stresses the necessity of an organized collagen matrix for proper bone formation.

B. Non-collagenous Proteins

Non-collagenous proteins, closely associated with bone collagen, are believed to play many of the roles once assigned to collagen, i.e., promoting the initial deposition of apatite mineral and regulating the size, orientation, and rate of growth of bone mineral crystals. These non-collagenous proteins account for 2 to 5% of the matrix of bone.

Establishing the role of these non-collagenous proteins in bone mineralization is an active area of research. The following sections review our current understanding of these roles.

1. Phosphoproteins

The phosphoproteins are a class of acidic proteins which are released from bone, dentin, and enamel during ethylenediamine tetracetic acid demineralization or guanadinium chloride dissociative extraction. Rich in aspartic acid and serine and containing high amounts of organic phosphate and dicarboxylic acids,[34] the phosphoproteins are found in elevated concentrations in areas where the mineral is well-oriented. Although they are not extractable from non-calcified cartilage, the phosphorus containing proteins account for approximately 42% of the non-collagenous proteins of newly formed subperiosteal calf bone.[35]

Because of their high affinity for calcium and/or apatite mineral, some or all of the bone phosphoproteins are thought to play a structural and/or regulatory role in hydroxyapatite deposition.[36] In vitro, phosphoproteins isolated from calcified tissues can (a) stabilize amorphous precursors of hydroxyapatite,[37] (b) cause hydroxyapatite to form directly in solutions, which due to their high supersaturations would normally form amorphous calcium phosphate,[37] (c) bind with high affinity to both collagen and hydroxyapatite[38] and (d) inhibit hydroxyapatite growth.[38]

Synthesis of phosphoproteins occurs coincident with, or after the start of, endochondral ossification.[39] This fact, plus the in vitro data described above, suggests that some of the phosphoproteins of bone may be involved in regulating the nucleation, orientation, and size of bone mineral crystals.

2. Glycoproteins

A large proportion of the proteins in bone are glycosylated. Some of these bone glycoproteins are essential for bone structure and organization. Some are involved in bone metabolism, while the function of others is yet to be determined. Bone collagen which contains 4 to 6% carbohydrate can be considered a glycoprotein. Of the non-collagenous proteins, several of the bone phosphoproteins discussed above are also glycoproteins. The bone morphogenic proteins which are believed to be responsible for bone induction are glycoproteins.[40] We still do not know the function of sialic acid containing glycoproteins which are found in bone.[41] Many of the enzymes involved in bone and cartilage metabolism, e.g., alkaline phosphatase, aryl sulfatase and some of the neutral and acid proteases, are glycosylated. The highly glycosylated α_2HS glycoprotein is the predominant non-collagenous bone protein during early embryonic life.[35] Another glycoprotein, isolated from subperiosteal bone, and found in relatively constant amount throughout fetal bone development[35] has been given the name osteonectin. A nectin is a protein which attaches to some chemical moiety. This glycoprotein accounts for 25% of fetal calf bone non-collagenous protein. Further, it is the only non-collagenous protein studied to date that has a high affinity in vitro for both hydroxyapatite and collagen.

Another glycoprotein found in bone is fibronectin.[42] This glycoprotein is involved in the regulation of cell-matrix and cell-cell interactions in all connective tissues. In contrast to osteonectin, fibronectin binds very tightly to collagen, but it does not associate directly with hydroxyapatite. Fibronectin plays a role in early bone development, and its presence is believed to be crucial for both wound healing and fracture repair.[10]

Autoradiography, using labeled glycoprotein precursors, demonstrates that glycoproteins are concentrated at the mineralization front in epiphyseal cartilage and present throughout bone.[43] In healing fracture callus, the pool of glycoproteins formed during early mineralization is constant in amount throughout all stages of bone formation.[44] The precise role of all the bone glycoproteins in mineralization as well as the nature of the interaction of bone glycoproteins with other matrix molecules is still under investigation.

3. GLA Proteins

Proteins containing gamma carboxy-glutamic acid (GLA), an anionic amino acid formed by a vitamin K-dependent, post-translational carboxylation of glutamic acid containing proteins, are major components of the non-collagenous proteins isolated from bone.[45,46] Absent from calcified cartilage and the entire epiphyseal plate, GLA-peptides first appear in cancellous bone and are concentrated in cortical bone. Density fractionation of compact bone reveals GLA-peptides associated with the densest, most heavily mineralized fractions. Bone GLA-protein, or osteocalcin, is the only bone protein whose in vitro synthesis is known to be regulated by vitamin D metabolites.[47] During endochondral bone induction,[48] base line levels of GLA are present during the early stages of bone formation, and bone GLA-protein increases in content only after bone remodelling commences, suggesting that the role of this protein may be to control the growth and/or remodelling of bone mineral.

4. Proteoglycans

Proteoglycans, which account for a major fraction on the organic matrix in non-calcified connective tissues are also components of the bone matrix. However, calcified cartilage and bone proteoglycans are quite different from those found in non-calcified cartilage. Non-calcified cartilage consists mainly of proteoglycan and collagen in a stiff, gel-like aqueous suspension. The high molecular weight, broadly-extended polyelectrolyte, proteoglycan, in combination with collagen and water produces a substance which is resistant to deformation, a viscoelastic property peculiar to cartilage. The proteoglycan aggregate molecule (MW 50 to 200 $\times$ 10^6 daltons) consists of a number of proteoglycan subunits bonded to a hyaluronic acid backbone by specific link glycoproteins (Figure 4). Each subunit (MW 1 to 4 $\times$ 10^6 daltons) is a bottle-brush arrangement of negatively charged glycosaminoglycans attached to a protein core. Both the aggregate and subunit molecules, because of the negative charges on the glycosaminoglycan chains, remain in a stretched out formation occupying a large volume of solution.

When cartilage calcifies, both the chemical nature and the distribution of proteoglycans are modified. Early histochemical and biochemical data showed a decrease in amount of proteoglycan at the growth plate mineralization front.[49] Thus, it has long been thought that proteoglycans are inhibitors of mineral formation which had to be modified and/or removed from mineralization to occur. Recent in vitro studies on the effect of bovine nasal proteo-

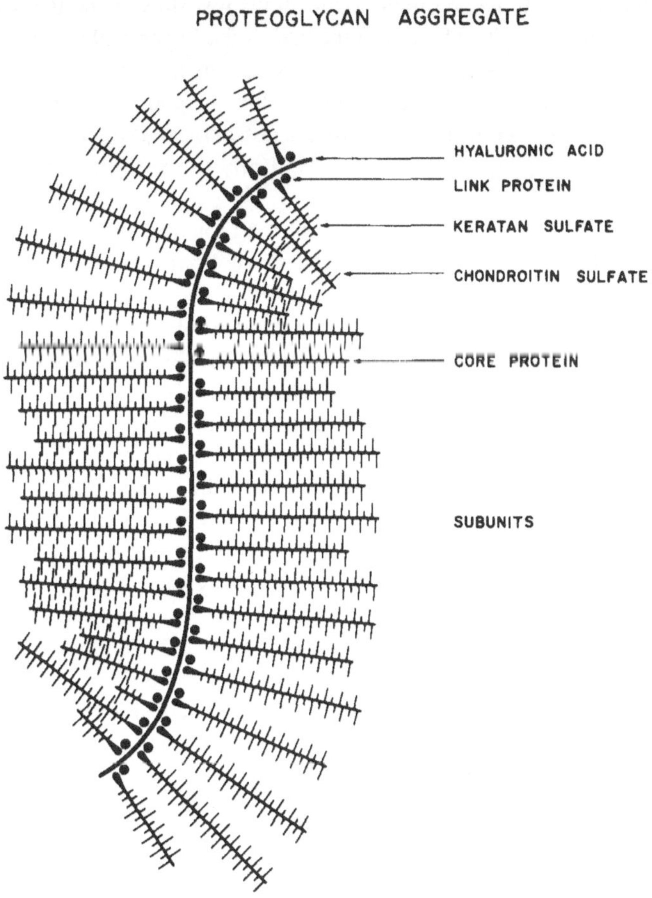

FIGURE 4. Diagram of proteoglycan aggregate molecule. (Courtesy of Dr. Lawrence C. Rosenberg. Reprinted by permission of Elsevier-North Holland Biomedical Press.)

glycan aggregates and subunits on hydroxyapatite formation support this view.[50] At concentrations found in non-calcifying cartilage, the aggregate and the subunit extracted from bovine nasal proteoglycan prevent hydroxyapatite precipitation from a supersaturated calcium phosphate solution. Proteoglycans appear to be less effective apatite inhibitors when sulfate groups are missing or when this macromolecule is modified in structure or reduced in molecular weight.[49]

During bone-matrix induced endochondral ossification[51] proteoglycan synthesis decreases markedly during cartilage calcification. During early bone formation there is an increase in synthesis of a low molecular weight proteoglycan. These low molecular weight proteoglycan aggregates have, on average, larger glycosaminoglycan subunits than cartilage proteoglycans

These data corroborate the thesis that proteoglycans must be modified for calcification to occur. The way these modifications take place and the mechanisms to explain how unmodified proteoglycans regulate mineral deposition remain to be determined.

C. Lipids and Carbohydrates

Bone contains lipids and carbohydrates which are involved in its formation and function. Many of the lipids of bone are components of cell and organelle membranes. Bone carbohydrates are found in proteoglycans, glycoproteins, and intracellular glycogen.

In endochondral ossification carbohydrate metabolism is slowed and the parallel synthesis of high energy nucleotide triphosphates diminished as the hypertrophic stage is approached. This is probably a result of decreased oxygen content in this area. A direct result of the shut-down of aerobic glycolysis is the release of calcium and phosphate from the mitochondria and subsequently from the cell. This in turn creates a local extracellular environment of elevated calcium and phosphate content which is ideally suited to support apatite nucleation and growth.

The lipids involved in tissue mineralization are the proteolipids and their component Ca-acidic phospholipid-PO_4 complexes.[49] The unique proteolipids in mineralizing tissues, found for the most part in membranes, are rich in acidic phospholipids. Such acidic phospholipids form Ca-acidic phospholipid-PO_4 complexes in calcium phosphate solutions; these complexes in turn promote hydroxyapatite deposition. The proteolipids in calcifying tissue are probably the major site for such complexes. Ca-acidic phospholipid-PO_4 complexes are present in highest concentrations at sites of rapid mineralization. In the bovine epiphyseal growth plate, the complexes are present in highest concentration in the lower half of the hypertrophic zone where mineralization begins.[52] In the healing fracture callus, the complex is at a maximum during the period of most rapid mineral deposition.[53] Similarly, in matrix-induced bone induction the complex is at its highest concentration at the start of mineral deposition. In contrast, these complexes are never present in non-calcifying soft tissue even though, as in the case of bone marrow, these tissues have a high total lipid content. The in vivo and in vitro studies on the association of Ca-acidic phospholipid-PO_4 complexes with apatite deposition suggest that these complexes, probably associated with membrane proteolipids, are among the factors which provide sites for initial mineral deposition.

In calcifying cartilage, it is likely that the Ca-acidic phospholipid-PO_4 complexes are part of the membranes of chondrocytes and extracellular matrix vesicles.[54] In membranous bone, where matrix vesicles are present,[55] the phospholipids may provide a site for mineral deposition. In normal cortical bone, where vesicles are not involved in hydroxyapatite deposition,[49] the site of lipid-induced mineral deposition is uncertain.

Lipids other than the acidic phospholipids discussed above are important for bone development serving as sources of nutrition and maintaining cellular integrity. Lipids in cell membranes control not only the flux of nutrients and electrolytes into and out of the cell, but also determine properties of membrane bound enzymes.

D. Enzymes and Hormones

The enzymes in bone can be classified as those which (a) are involved in the biosynthesis of matrix components, (b) are needed for remodeling of bone, and (c) are specifically concerned with mineral deposition. In the first group are those enzymes needed for translational and post-translational modification of matrix proteins. The most abundant remodeling, or degradative, enzymes are the metalloproteases, such as collagenase and proteoglycanase, and the lysosomal enzymes, such as the cathepsins and acid phosphatase. Enzymes concerned with mineral deposition include: alkaline phosphatases, inorganic pyrophosphatase, and ATPases. These membrane-bound enzymes are thought to regulate the elevation of the local phosphate concentration in calcifying fluids thereby causing mineral deposition at a desired location. The reader is referred to Vaughan[56] for a more detailed picture of the role of enzymes in hard tissues.

The role of hormones in bone structure and function will not be dealt with in this review. However, it is important to note that all the processes occurring in bone are controlled by hormone action. For example, vitamin D and its metabolites regulate the transport of calcium and phosphate into and out of bone. Parathyroid hormone and calcitonin are antagonistically involved in controlling plasma calcium levels. Growth hormones and corticosteroids affect bone development as a result of their effect on protein synthesis and degradation.

FIGURE 5. Cartoon illustrating the opposing forces active at the mineralization front.

IV. CONCLUSIONS

The mineral and organic constituents of bone interact throughout bone development to give each type of bone its characteristic properties. Certain bones, such as the long bones, grow through a calcified cartilage precursor. Other bones, such as the flat bones, form directly by mineralizing the extracellular matrix without going through a calcified cartilage stage. In the mineralization of all bones three processes are possible: (1) the local raising of the calcium phosphate supersaturation, (2) the removal or change in inhibitors to calcification, and (3) the active participation of certain apatite nucleators. For example, an enzyme such as alkaline phosphatase could raise the local phosphate concentration (and the calcium phosphate supersaturation) by cleaving phosphate ester substrates. Proteoglycan inhibitors could be modified or degraded before mineralization occurs. Proteolipids and certain phosphoproteins might serve as specific apatite nucleators.

The precise interplay between the many factors involved in tissue mineralization is not known. Other factors, yet to be discovered, may play significant roles in preparing the tissue for mineral deposition and in controlling the growth and remodeling of bone mineral. There is not a unique apatite inhibitor or nucleator, so that the controls on mineralization can be considered redundant. Figure 5 illustrates this redundancy with a cartoon showing the opposing forces promoting and retarding mineral deposition at the mineralization front. It should be clear from this review that more work is needed on the isolation and characterization of the many molecules and factors involved before we can see a complete picture of how mineralization of bone is accomplished.

ACKNOWLEDGMENTS

This is Publication Number 152 of the Laboratory of Ultrastructural Biochemistry. The experimental work cited in this review was supported in part by NIH Grant DE-04141, AM-18412 and AM-19776. Our thanks to Dr. Peter G. Bullough for his many useful suggestions including the ideas incorporated in Figure 5.

REFERENCES

1. **Ham, A. W. and Cormack, D. W.,** *Histology*, J.B. Lippincott, Philadelphia, 1979, 377.
2. **Kember, N. F. and Sissons, H. A.,** Quantitative histology of the human growth plate, *J. Bone J. Surg.*, 58B, 426, 1976.
3. **Trueta, J.,** The role of the vessels in osteogenesis, *J. Bone J. Surg.*, 45B, 402, 1963.
4. **Brighton, C. T. and Hunt, R. M.,** Mitochondrial calcium and its role in calcification. Histochemical localization of calcium in electron micrographs of the epiphyseal growth plate with K-pyroantimonate, *Clin. Orthop.*, 100, 406, 1974.
5. **Posner, A. S.,** Intramitochondrial storage of stable amorphous calcium phosphate, *Ann. N.Y. Acad. Sci.*, 307, 248, 1978.
6. **Anderson, H. C.,** Vesicles associated with calcification in the matrix of epiphyseal cartilage, *J. Cell. Biol.*, 41, 59, 1969.
7. **Anderson, H. C.,** Calcification Processes, *Path. Ann.*, 15, 45, 1980.
8. **Boskey, A. L.,** Models of matrix vesicle calcification, *Inorg. Persp. Biol. Med.*, 2, 51, 1979.
9. **Udupa, K. M. and Prasad, G. C.,** Chemical and histochemical studies on the organic constituents in fracture repair in rats, *J. Bone J. Surg.*, 45B, 770, 1963.
10. **Reddi, A. H.,** Cell biology and biochemistry of endochondral bone development, *Coll. Res.*, 1, 209, 1981.
11. **Kay, M. I., Young, R. A., and Posner, A. S.,** The crystal structure of hydroxyapatite, *Nature*, 204, 1050, 1964.
12. **Montel, G.,** Conceptions nouvelles sur la physico-chemie des phosphates de structure apatitique, *Bull. Soc. Chim.*, Special No., 1693, 1968.
13. **LeGeros, R. Z., Trautz, O., LeGeros, J. P., and Klein, E.,** Carbonate substitution in the apatite structure, *Bull. Soc. Chem. Fr.*, Special No., 1712, 1968.
14. **Blumenthal, N. C., Betts, F., and Posner, A. S.,** Effect of carbonate and biological macromolecules on formation and properties of hydroxyapatite, *Calc. Tiss. Res.*, 18, 81, 1975.
15. **Blumenthal, N. C. and Posner, A. S.,** Hydroxyapatite: mechanism of formation and properties, *Calc. Tiss. Res.*, 13, 235, 1973.
16. **Boskey, A. L. and Posner, A. S.,** Formation of hydroxyapatite at low supersaturation, *J. Phys. Chem.*, 80, 40, 1976.
17. **Landis, W. J., Paine, M. C., and Glimcher, M. J.,** Considerations for the electron optical identification of matrix vesicles and mineral phase particles possibly associated with them in calcifying tissues, *Trans. Orthopaedic Res. Soc.*, 6, 59, 1981.
18. **Wuthier, R. E. and Eanes, E. D.,** Effect of phospholipids on the transformation of amorphous calcium phosphate to hydroxyapatite in vitro, *Calc. Tiss. Res.*, 23, 135, 1975.
19. **Ali, S. Y., Wisby, A., and Gray, J. C.,** Electron probe analysis of cryosections of epiphyseal cartilage, *Met. Bone Dis. Rel. Res.*, 1, 97, 1978.
20. **Murphree, S., Hsu, H. H. T., and Anderson, H. C.,** In vitro formation of crystalline apatite by matrix vesicles isolated from rachitic rat epiphyseal cartilage, *Calcif. Tiss. Int.*, 34, 562, 1982.
21. **Herzfeld, J., Roufosse, A., Haberkorn, R. A., Griffin, R. G., and Glimcher, M. J.,** Magic angle sample spinning in inhomogeneously broadened biological systems, *Phil. Trans. R. Soc. Lond.*, B 289, 459, 1980.
22. **Berry, E. D.,** The structure and composition of some calcium deficient apatites, *J. Inorg. Nucl. Chem.*, 29, 317, 1967.
23. **Meyer, J. L.,** Hydroxyl content of solution precipitated calcium phosphates, *Calc. Tiss. Int.*, 27, 153, 1979.
24. **Posner, A. S.,** The crystal chemistry of bone mineral, *Physiol. Revs.*, 49, 760, 1969.
25. **Francis, M. D., Briner, W. W., and Grey, J. A.,** Chemical agents in the control of calcification processes in biological systems, in *Hard Tissue Growth, Repair and Remineralization*, Sognnaes, R. F. and Vaughan, J. M., Eds., Elsevier, Amsterdam, 1973, 57.
26. **Jackson, S. A., Cartwright, A. S., and Lewis, D.,** The morphology of bone mineral crystals, *Calc. Tiss. Res.*, 25, 217, 1978.
27. **Eyre, D.,** Collagen: Molecular diversity in the body's protein scaffold, *Science*, 207, 1315, 1980.
28. **Gay, S., Muller, P. K., Lemmen, C., Remberger, K., Matzen, K., and Kuhn, K.,** Immunohistological study on collagen in cartilage bone metamorphosis and degenerative osteoarthritis, *Klin. Wschr.*, 54, 969, 1976.
29. **Von der Mark, K. and Von der Mark, H.,** The role of three genetically distinct collagen types in endochondral ossification and calcification of cartilage, *J. Bone J. Surg.*, 59B, 458, 1977.
30. **Reddi, A. H., Gay, R., Gay, S., and Miller, E. J.,** Transitions in collagen types during matrix-induced cartilage, bone and bone marrow formation, *Proc. Nat. Acad. Sci.*, 74, 5589, 1977.
31. **White, S. W., Holmes, D. J. S., Miller, A., and Timmins, P. A.,** Collagen-mineral axial relationship in calcified turkey leg tendon by X-ray and neutron diffraction, *Nature*, 266, 421, 1977.

32. **Glimcher, M. J.,** Composition, structure and organization of bone and other mineralized tissues and the mechanism of calcification, in *Handbook of Physiology-Endocrinology VII*, Williams & Wilkins, Baltimore, 1976, 25.
33. **Levene, D.,** Diseases of the collagen molecule, *J. Clin. Path.*, 319, 82, 1979.
34. **Lee, S. L. and Glimcher, M. J.,** Purification, composition and ^{31}P-NMR spectroscopic properties of a non-collagenous phosphoprotein isolated from chick bone matrix, *Calcif. Tissue Int.*, 33, 385, 1981.
35. **Termine, J. D., Belcourt, A. B., Conn, K. M., and Kleinman, H. K.,** Mineral and collagen binding proteins of fetal calf bone, *J. Biol. Chem.*, 256, 10403, 1981.
36. **Veis, A.,** The role of acidic proteins in biological mineralization, in *Ions in Macromolecular and Biological Systems*, D. H. Everett and B. Vincent, Eds., Scietechnia, Bristol, 1978, 259.
37. **Nawrot, C. F., Campbell, D. J., Schroeder, J. K., and van Valkenberg, M.,** Dental phosphoprotein-induced formation of hydroxyapatite during in vitro synthesis of amorphous calcium phosphate, *Biochem. J.*, 5, 3445, 1976.
38. **Termine, J. D.,** Bone and tooth mineralization: matrix effects and crystal development, *Prog. Cryst. Growth Charact.*, 3, 65, 1980.
39. **Glimcher, M. J., Kossiva, D., and Roufosse, A.,** Identification of phosphopeptides and γ-carboxyglutamic acid-containing peptides in epiphyseal growth plate cartilage, *Calcif. Tiss. Int.*, 27, 187, 1979.
40. **Urist, M. R. and Mikulski, A. J.,** A soluble bone morphogenic protein extracted from bone matrix with a mixed aqueous and non-aqueous solvent, *Proc. Soc. Exp. Biol. Med.*, 162, 48, 1979.
41. **Herring, G. M. and Kent, P. W.,** Some studies on muco substances of bovine cortical bone, *Biochem. J.*, 89, 405, 1963.
42. **Pearlstein, E., Gold, L., and Garcia-Pardo, A.,** Fibronectin: A review of its structure and biological activity, *Mol. Cell. Biochem.*, 29, 103, 1980.
43. **Weinstock, M.,** Radioautographic visualization of ^{3}H-fucose incorporation into glycoprotein by osteoblasts and its deposition into bone matrix, *Calcif. Tiss. Int.*, 27, 177, 1979.
44. **Lane, J. M., Golembiewski, G., Boskey, A. L., and Posner, A. S.,** Biochemical studies of the fracture callus matrix: Effect of mode of orthopedic stabilization on estimated proteoglycan contents, *Met. Bone Dis. Rel. Res.*, 4, 61, 1982.
45. **Hauschka, P.,** Osteocalcin in developing bone systems, in *Vitamin K Metabolism and Vitamin K Dependent Protein*, J. W. Suttie, Ed., University Park Press, Baltimore, 1979, 277.
46. **Price, P. A., Otsuka, A. S., Posey, J. E., Kristaponis, J., and Raman, N.,** Characterization of a γ-carboxyglutamic acid-containing protein from bone, *Proc. Nat. Acad. Sci.*, 73, 1447, 1976.
47. **Price, P. A. and Baukel, S. A.,** 1,25-dihydroxyvitamin D_3 increases synthesis of the Vitamin K-dependent bone protein by osteosarcoma cells, *J. Biol. Chem.*, 255, 11660, 1980.
48. **Price, P. A., Lothringer, J. W., Baukel, S. A., and Reddi, A. H.,** Developmental appearance of the vitamin K-dependent protein of bone during calcification, *J. Biol. Chem.*, 256, 3781, 1981.
49. **Boskey, A. L.,** Current concepts of the physiology and biochemistry of calcification, *Clin. Orthoped.*, 157, 225, 1981.
50. **Blumenthal, N. C., Posner, A. S., Silverman, L. S., and Rosenberg, L. C.,** The effect of proteoglycans on in vitro hydroxyapatite formation, *Calcif. Tiss. Int.*, 27, 75, 1979.
51. **Reddi, A. H., Hascall, V. C., and Hascall, G. K.,** Changes in proteoglycan type during matrix-induced cartilage and bone development, *J. Biol. Chem.*, 253, 2429, 1978.
52. **Boskey, A. L., Posner, A. S., Lane, J. M., Goldberg, M. R., and Cordella, D. M.,** Distribution of lipids associated with mineralization in the bovine epiphyseal growth plate, *Arch. Biochem. Biophys.*, 199, 305, 1980.
53. **Boskey, A. L., Timchak, D. M., Lane, J. M., and Posner, A. S.,** Phospholipid changes during fracture healing, *Proc. Soc. Exp. Biol. Med.*, 165, 368, 1980.
54. **Wuthier, R. E. and Gore, S. T.,** Partition of inorganic ions and phospholipids in isolated cell, membrane and matrix vesicle fractions: Evidence for Ca-Pi-acidic phospholipid complexes, *Calcif. Tiss. Res.*, 24, 163, 1977.
55. **Bernard, G. W. and Pease, P. C.,** An electron microscopic study of initial intramembranous osteogenesis, *Am. J. Anat.*, 125, 271, 1969.
56. **Vaughan, J.,** *The Physiology of Bone*, 3rd ed., Clarendon Press, Oxford, 1981, 208.
57. **Pugliarello, M. C., Vuttur, F., and de Bernard, B.,** Analysis of bone composition at the microscopic level, *Calcif. Tiss. Res.*, 12, 209, 1973.
58. **Herring, G. M.,** Methods for the study of the glycoproteins and proteoglycans of bone using bacterial collagenase. Determination of bone sialoprotein and chondroitin sulphate, *Calcif. Tiss. Res.*, 24, 29, 1977.
59. **Vejlens, L.,** Glycosaminoglycans of human bone tissue, I. Pattern of compact bone in relation to age, *Calcif. Tiss. Res.*, 7, 175, 1971.
60. **Pellegrino, E. D. and Biltz, R. M.,** The composition of human bone in uremia, *Medicine*, 44, 397, 1965.

Chapter 4

ELASTICITY AND VISCOELASTICITY OF CORTICAL BONE

W. Bonfield

TABLE OF CONTENTS

I. INTRODUCTION

The specific topics of elasticity and viscoelasticity of compact or cortical bone have been featured in a variety of more comprehensive reviews of the mechanical properties of bone.[1,2] Consequently, the objective of this chapter is not to repeat a catalog of all the various results, but instead to consider the overall implications of selected data with respect to a current definition and understanding of elastic and viscoelastic deformation of bone. There is an obvious connection between this topic and the concept of yielding in bone, which is dealt with elsewhere in this volume, and to which the reader is referred.

II. MACROSCOPIC STRESS-STRAIN BEHAVIOR

A schematic illustration of typical load-deformation (or stress-strain) curves for regular sections of compact bone,[3] tested in tension or compression, in the "wet" condition (i.e. soaked in Ringer's solution or water), at room temperature (say 18 to 25°C) is shown in Figure 1. The curves are of two types: first, an apparently straight line to fracture, and second, an apparently straight line followed by a decreasing strain hardening rate to fracture. The transition in slope in the second curve has been generally associated with yielding, namely the onset of non-elastic deformation, but it could also be considered as the onset of micro-cracking. However, as even in this case the extent of post yield strain is limited, with a maximum total elongation of only 0.5%[3] to 3%[4], bone may be generally classified as a brittle rather than a ductile solid. Consequently, an evaluation of the fracture of bone is conditioned by the nature of any internal or surface "space" provided by microstructural features such as Haversian canals, lacunae, etc., or in particular, by defects such as notches or cracks, (the latter perhaps introduced by specimen preparation), which lead to a wide variation in the measured fracture stress. Hence, fracture characterization in bone requires an appropriate fracture mechanics treatment to account for both the applied stress and the nature of the critical crack (e.g. see Reference 5). The effect of preparation-induced cracks is less important in a consideration of the deformation of bone as they produce a variation in the extent, but not in the slope of the "straight" line portion of the stress-strain curves in Figure 1, which, in many investigations, has been taken to indicate that bone acts as a linear elastic solid in this region of the stress-strain curve. Hence the linear slope of the stress (σ)-strain (ϵ) curve provides a direct measure of the Young's modulus (E) of bone, with

$$E = \sigma/\epsilon \qquad (1)$$

As bone is a heterogeneous, multiphase, composite, E does not have a unique value, but varies with orientation, with a maximum value parallel to the long axis of a support bone, such as the tibia, and a minimum value, approximately a factor of ~ 2 smaller, perpendicular to the long axis, such as demonstrated by Dempster and Liddicoat.[6] The absolute values measured for E depend on the precise nature of the bone specimen and its microstructure, but normally are within a range of 7 to 30 GNm^{-2} (Reference 7) for the longitudinal orientation of mature compact bone in the "wet" condition. It is interesting to note in the context of prosthetic materials used in combination with bone, that the values of E for bone lie between those for polymers (e.g. polyethylene (~ 1 GNm^{-2})) and alloys, (e.g. Ti 6% Al 4% V (106 GNm^{-2}), stainless steel (200 GNm^{-2})). The variation of E with orientation for bovine compact bone shows a non-uniform decrease[8] from the longitudinal (0°) to the transverse orientation (90°), as illustrated in Figure 2. These data do not follow the predictions of a model of bone as a fiber-reinforced composite (i.e. hydroxyapatite-reinforced collagen), a point which is returned to in Section V. The relevant data points in Figure 2 agree reasonably

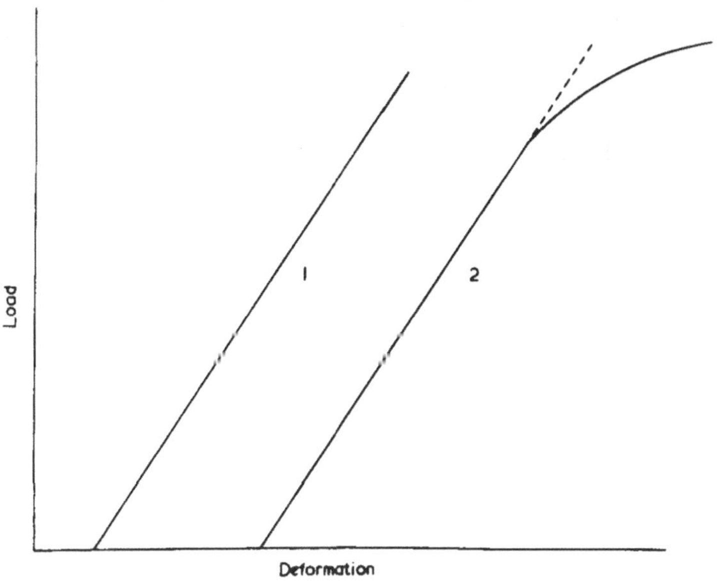

FIGURE 1. Schematic illustration of load-deformation (or stress-strain) curves obtained for cortical bone specimens. (From Bonfield, W. and O'Connor, P., *J. Mater. Sci.*, 13, 202, 1978. With permission.)

with the values obtained by Reilly and Burstein[9] for human compact bone with E values at 0°, 30°, 60° and 90° of 23.1, 16.7, 12.8, and 10.4 GNm^{-2}, respectively. However, it should be emphasized that the microstructural complexity of bone requires more data at orientations other than 0° and 90° before the experimental variation of E with orientation can be precisely established.

The value of E also depends on the test temperature and decreases in a linear manner with an increase in temperature in the range from 17° to 41°C, as shown in Figure 3.[10] It should be noted that the values measured in this temperature range and indeed in the range from −58°C to 41°C are reversible, i.e. they are not affected by previous tests at other temperatures. The critical temperature is of course the physiological temperature of 37°C, at which tests are not generally performed and at which E is less than the room temperature value (Figure 3). Tests at a temperature of ~ 90°C produce a substantial reduction in E and give a different type of stress-strain behavior.[23]

The Young's modulus of bone also depends on the strain rate of the test, an effect originally demonstrated in a convincing manner, for the longitudinal orientation in compression, by McElhaney,[11] as shown in Table 1 and since confirmed by many workers (e.g. References 12-14), with the comprehensive study of Wright and Hayes[15] in tension, establishing an increase of E by a factor of ~ 2 (i.e. ~ 17 to 40 GNm^{-2}) with an increase in strain rate from the "quasi-static" range, e.g. $5 \times 10^{-4}\ s^{-1}$ to $2.4 \times 10^{2}\ s^{-1}$. As discussed by Wright and Hayes, an increase of E with strain rate of this magnitude appears to represent most of the data gathered in various investigations, either as a function of strain rate or at particular strain rates. It should be noted that there is an almost complete absence of E values outside the "quasi-static" strain rate range for bone orientations other than the longitudinal orientation. However one result for the transverse orientation obtained during loading in a shock tube[16] gave E = 13.2 GNm^{-2} at a strain rate of 7 s^{-1}, compared with E ~10 GNm^{-2} at $10^{-3}\ s^{-1}$ (Reference 9), which suggests that a similar Young's modulus-strain rate relationship prevails.

The preceding discussion has referred generally to bone in the "mature" condition, in which the particular age does not appear as a significant or reproducible factor. However,

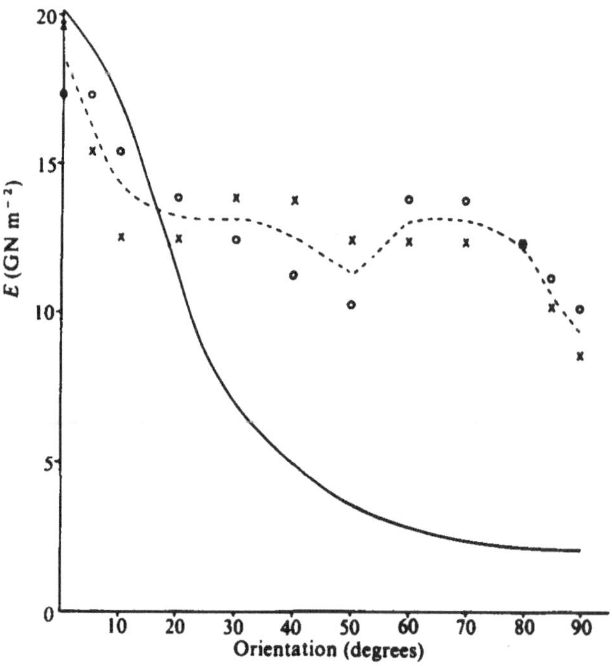

FIGURE 2. Variation in Young's modulus (E) of bovine femur specimens with the orientation of the specimen axis to the long axis of the bone, for wet (○) and dry (x) conditions, compared with the theoretical curve predicted from a fiber-reinforced composite model (From Bonfield, W. and Grynpas, M. D., *Nature*, 270, 453, 1976. Copyright Macmillan Journals. With permission.)

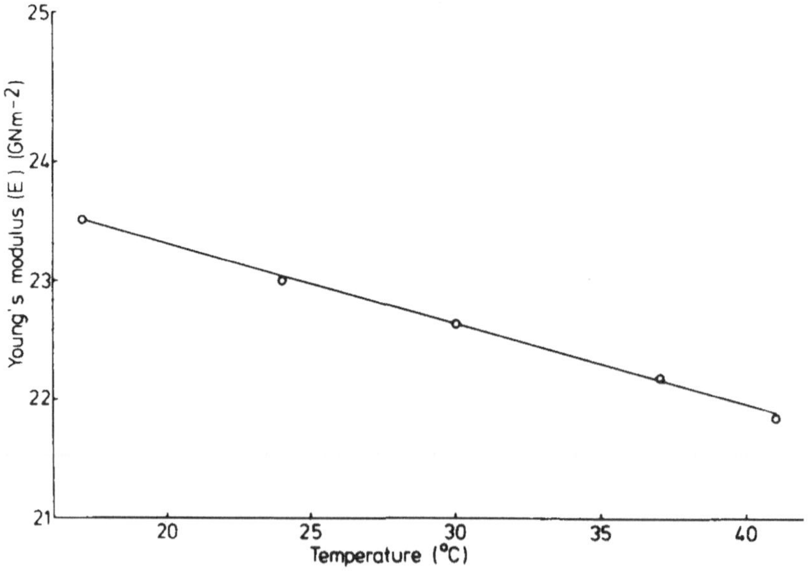

FIGURE 3. The dependence of the average longitudinal Young's modulus of bovine femur cortical bone on test temperature. (From Bonfield, W. and Tully, A., *J. Biomed. Eng.*, 4, 23, 1982, by permission of the publishers, Butterworth & Co.)

Table 1
EFFECT OF STRAIN RATE ON THE YOUNG'S MODULUS OF BOVINE COMPACT BONE[11]

Strain rate (s^{-1})	Young's modulus (GNm^{-2})
0.001	18.6
0.01	20.0
0.1	24.4
1.0	27.6
300	33.1
1500	42.0

during growth to the mature state there is a substantial increase in E with age, associated largely with the combined effects of an increase in hydroxyapatite volume fraction and a decrease in vascular space. For example, it was demonstrated[13] that maturation in rabbit tibiae was accompanied by an increase in E from 15.1 (at 3 months) to 27.6 GNm^{-1} (at 12 months or greater).

A recent comprehensive study by Torzilli et al.[17] has convincingly documented the increase in E during early growth of canine bone by a combination of whole bone and section tests, as shown in Figure 4. It can be seen from Figure 4 that the Young's modulus increases dramatically with age, from 0.2 to 1.2 GNm^{-2} (i.e., values approaching that of collagen) to ~ 20 GNm^{-2}, during the first 24 weeks, following which there was no significant age-related change. However, it should be noted that differences in E exist at any given age between different bones, which reflect the corresponding differences in bone microstructure. Moreover, even for similar bone, or within a given bone, differences in E exist, which reflect local structural variations. This effect was demonstrated by Currey,[18] who found a sensitive dependence of E (for the longitudinal direction) in "similar" mature rabbit bone on ash content (which provides a measure of the hydroxyapatite content) as shown in Figure 5. It can be seen that E increases with relatively small increases in ash content, although the extent of the range from ~ 6 to ~ 17 GNm^{-2} is greater than would be expected if hydroxyapatite volume fraction was the sole controlling factor (see Section V). For example, Abendschein and Hyatt[19] found a smaller range of E values, between ~ 17 to ~ 21 GNm^{-2}, for human cortical bone sections within the normal density variation of ~ 1.85 to 2.05 Mgm^{-3} (Figure 6). An increase in bone density for normal bone would be associated with an increase in hydroxyapatite (the denser phase) relative to collagen in bone, but would also reflect any decrease in vascular space within bone. One such demonstration of the possible contribution of vascular space is provided by Abendschein and Hyatt's[19] experiments on pathological bone, in which E values of ~ 8 GNm^{-2} were recorded for the abnormally low density values of ~ 1.6 Mgm^{-3}, as shown in Figure 6. While it is apparent that the Young's modulus of bone does depend on its ash content and/or density, the relationship between these parameters and the details of bone microstructure (i.e. proportion of secondary osteons, osteon size, hydroxyapatite orientation, etc.), remains to be established — indeed, the complexity of the microstructure as revealed by elegant structural characteriztion of Ascenzi (e.g. Reference 20), makes this a daunting task.

III. MICROSCOPIC STRESS-STRAIN BEHAVIOR

If the tests illustrated for mature bone in Figure 1 were stopped within the straight line portion of the graph, and the specimen unloaded to zero stress, then it might be expected

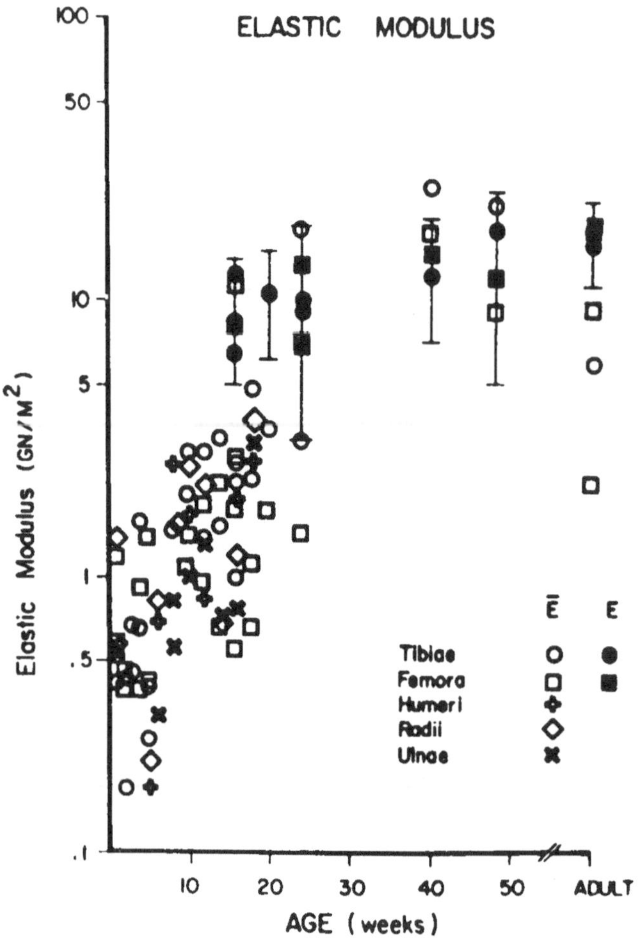

FIGURE 4. Canine bone elastic (Young's) modulus determined from whole bone (Ē) and bone section (E) tests. (From Torzilli, P. A. et al., *Mechanical Properties of Bone,* AMD, Vol. 45, American Society of Mechanical Engineers, New York, 1981. With permission.)

that the *unloading* curve would coincide with the loading curve, i.e. the behavior was simply linear elastic. This would contrast with the effect of unloading from a stress in the post-yield region of the second curve in Figure 1, in which it would be expected that the unloading curve would not correspond to the loading curve, i.e. an irreversible change had been produced by non-elastic deformation. In fact, the situation in bone is more complicated than this model, a finding which was obscured for many years because of two factors: first, many investigators only tested bone on loading and, second, even if the specimen was subsequently unloaded, the strain measurement sensitivity was not sufficient to distinguish any significant difference between the loading and unloading curves in the elastic region. In 1966 to 1968, Bonfield and Li[21-23] demonstrated with the use of high-sensitivity microstrain measurements that above a critical and small stress, which was considerably below the level associated with the *macroscopic* yielding depicted in Figure 1, the load-unload curves did not coincide, the unloading slope was smaller than the loading slope, and a residual strain was always present on first unloading to zero stress, which subsequently recovered with time. In these first experiments, it was suggested that some residual strain remained after recovery, i.e. plastic or permanent strain was produced, but later work[3,7] amply demonstrated that the residual strain was completely recovered if sufficient time was allowed for the process to

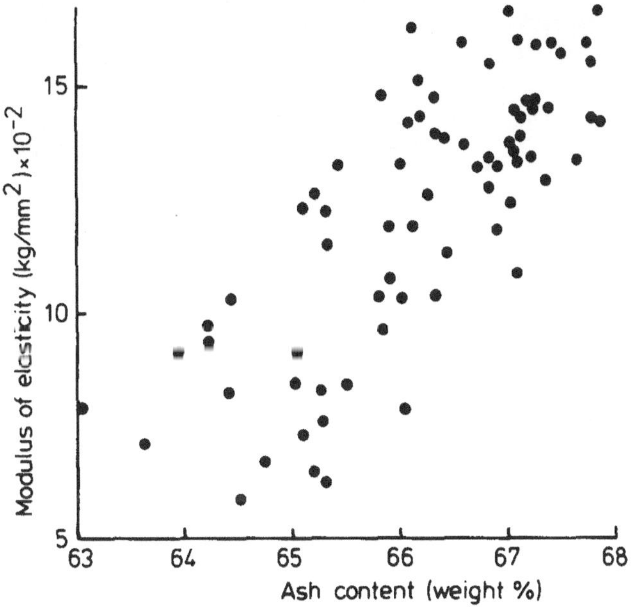

FIGURE 5. Relationship between the modulus of elasticity (Young's modulus) of rabbit cortical bone and ash content. (From Currey, J. D., *J. Biomech.*, 2, 1, 1969. With permission.)

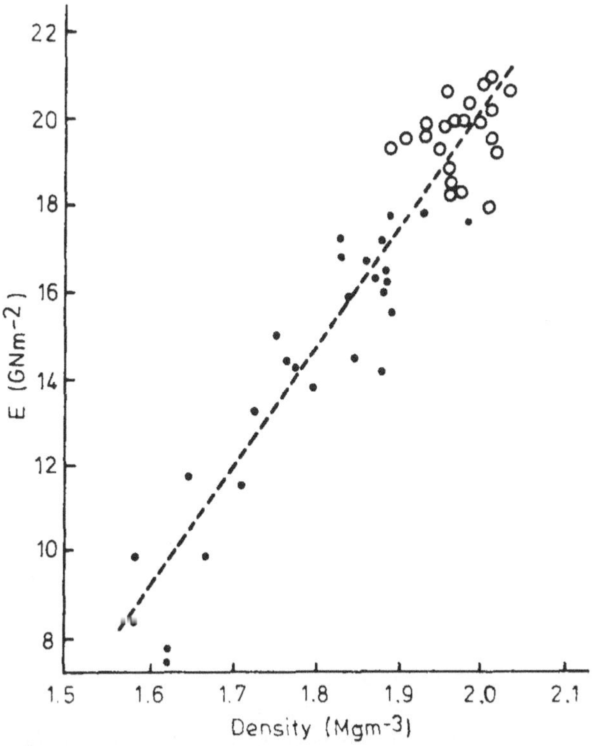

FIGURE 6. Relationship between the Young's modulus (E) of human bone and density for normal (○) and pathological specimens (●). (From Abendschein, W. G. and Hyatt, G. W., Ultrasonic and selected physical properties of bone, *Clin. Orthop.*, 69, 294, 1970. With permission.)

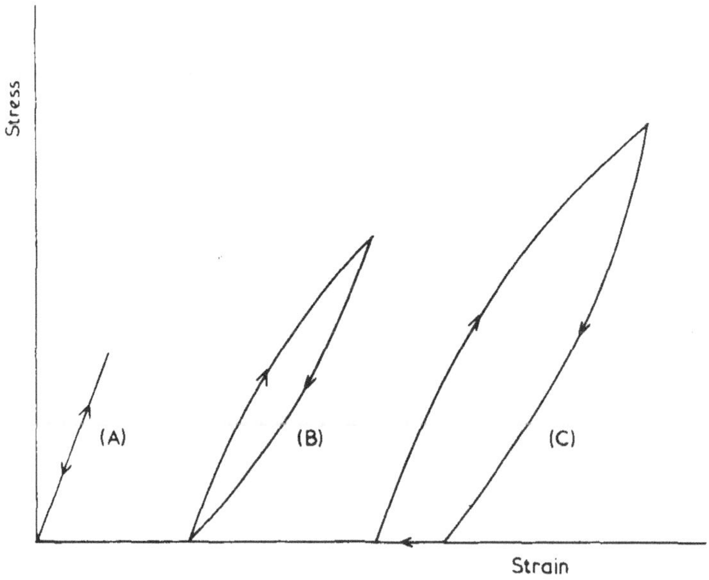

FIGURE 7. Schematic illustration of the three types of load-unload behavior observed for cortical bone, with (A) a straight line (linear elastic), (B) a closed hysteresis loop (anelastic) and (C) an initially open loop, which closes with time at zero stress (also anelastic). (From Bonfield, W. and O'Connor, P., *J. Mater. Sci.*, 13, 202, 1978. With permission.)

take place — i.e. anelastic or viscoelastic strain (recoverable with time) was produced. A schematic illustration of this process is shown in Figure 7 and illustrates that the linear elastic region (i.e. the straight line A) is followed by the hysteresis loops (B and C) produced by a combination of elastic *and* anelastic or viscoelastic deformation. Hysteresis loop B represents the situation when the load-unload curves coincide at zero stress, whereas hysteresis loop C illustrates the more general case, when an initial residual strain was obtained on unloading, *which recovered with time* at zero stress to form a *closed* loop. Typical recovery curves are shown in Figure 8. The stress level required to produce an initial residual strain of 2×10^{-6}, which represented the experimental strain measurement sensitivity, and defined as the *microscopic* yield stress (MYS), was 4 to 20 MNm^{-2}.

One consequence of these findings is that measurements of E should ideally take account of the viscoelastic component, although the effect is relatively small, as is shown in a microstrain representation of an "apparently" straight stress-strain curve (Figure 9). A more important conclusion is that hysteresis loops will be produced during the strain levels reached in physiological loading.[24] Moreover, the loop amplitude can be decreased, and finally eliminated, if the specimen is not allowed to recover completely between loading cycles,[22] an effect which could have important consequences in fatigue.[25] Hysteresis loops are produced by reversible dissipation of energy during the loading-unloading cycle. This effect has been a subject of considerable interest in research on the microplasticity of metals; Bonfield and O'Connor[3] utilized such an approach to investigate the relationship between the energy dissipated, as represented by the loop area (ΔW), and the strain amplitude, as given by the loop width ($\Delta\gamma$), as shown in Figure 10 for longitudinal bovine compact bone specimens. It can be seen that there is a reasonable linear relationship between ΔW and $\Delta\gamma$, from which a critical stress for the onset of anelastic deformation, defined as the friction stress (σ_F), was obtained[3] from:

$$\sigma_F = \frac{\Delta W}{2\Delta\gamma} \tag{2}$$

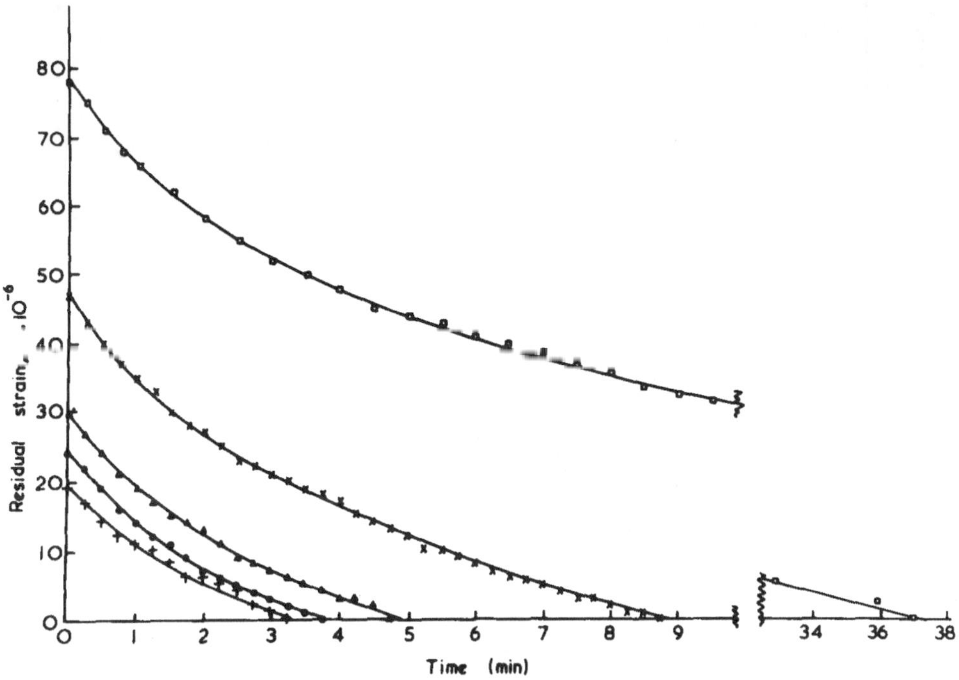

FIGURE 8. Recovery of residual non-elastic strain for bovine femur cortical bone with time at zero stress following various prestrains, (□) 3.01 × 10^{-3}, (x) 2.81 × 10^{-3}, (▲) 2.60 × 10^{-3}, (○) 2.21 × 10^{-3}, (+) 2.00 × 10^{-3}. (From Bonfield, W. and O'Connor, P., *J. Mater. Sci.*, 13, 202, 1978. With permission.)

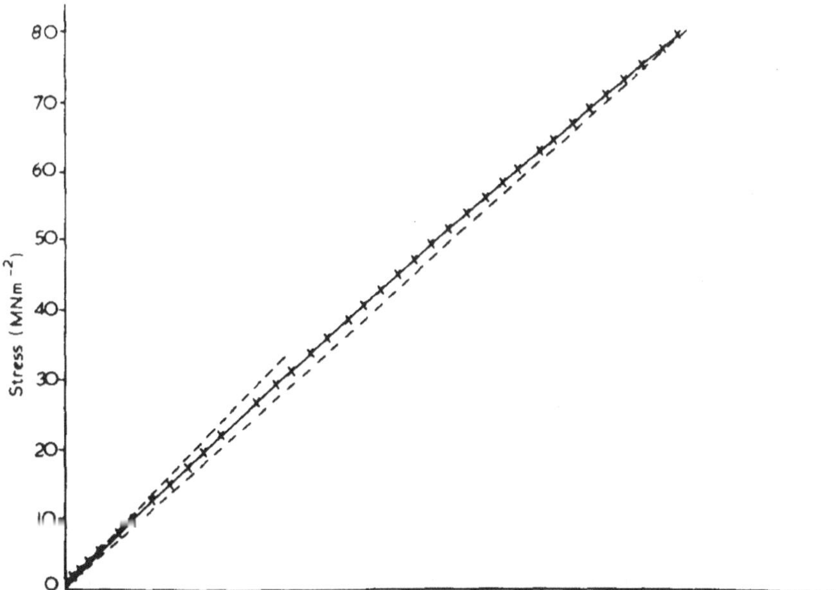

FIGURE 9. Stress — total strain behavior to fracture for a bovine femur cortical bone section as determined with microstrain measurements. (From Bonfield, W. and O'Connor, P., *J. Mater. Sci.*, 13, 202, 1978. With permission.)

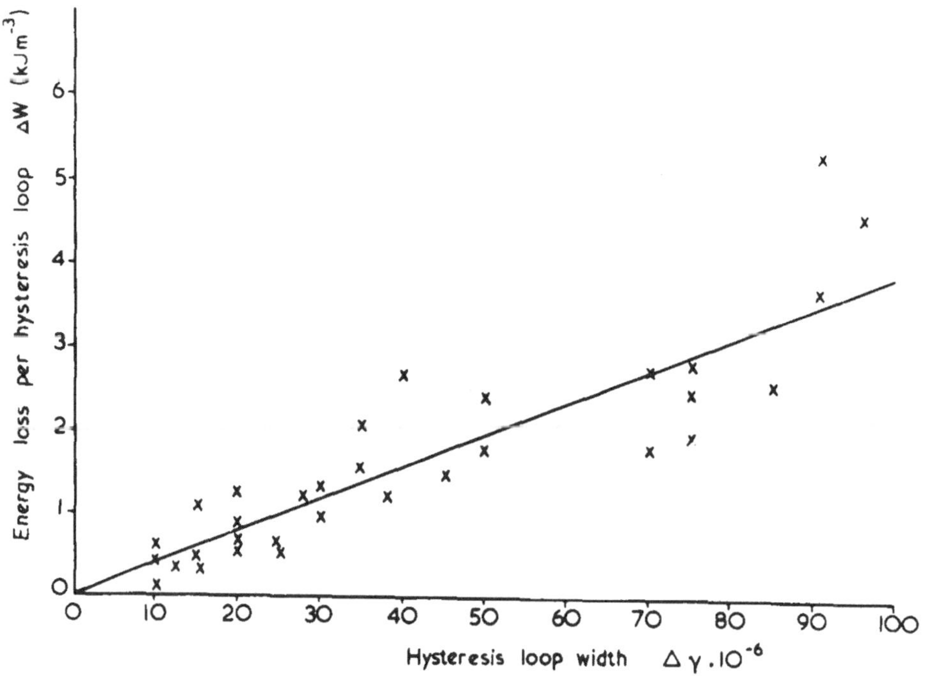

FIGURE 10. The variation of energy loss per hysteresis loop with forward non-elastic strain amplitude for bovine femur cortical bone sections. (From Bonfield, W. and O'Connor, P., *J. Mater. Sci.*, 13, 202, 1978. With permission.)

which for Figure 10, gave $\sigma_F = 19.8$ MNm^{-2}. While this analysis requires further development, it is encouraging to note that the value of σ_F derived indirectly is comparable to the direct measurement value of MYS, although it should be emphasized that σ_F and MYS have a different physical interpretation.

The amount of viscoelasticity produced by a given stress application increases if the strain rate is decreased. The lower limit to a decrease in strain rate is provided by a creep test for which, as was first demonstrated by Currey,[26] considerable anelastic strain can be produced. There does appear to be an upper limit to the magnitude of anelastic strain, as O'Connor[27] found that specimens loaded at ~ half the fracture stress would creep to fracture in ~ 24 hr. The establishment that viscoelasticity was significant in "quasi-static" tests is paralleled by investigations under dynamic conditions,[2] which have demonstrated the complex nature of the controlling mechanisms.

IV. ELASTIC CONSTANTS

The preceding discussion of the parameters controlling deformation has interpreted elasticity in bone simply in terms of Young's modulus (E), which represents one of the technical elastic constants. A more complete description can be obtained by reference also to the shear modulus (G) and Poisson's ratio (ν). As bone is anisotropic, the technical elastic constants would ideally be related to specific directions in bone, but this raises the fundamental question as to the symmetry of bone. For example, if it is assumed that bone is a linear elastic solid which is transversely isotropic, then there are 12 non-zero components in the stiffness matrix, of which *five* are independent and require measurement. This approach was followed successfully by Lang,[28] using an ultrasound technique, and Reilly and Burstein,[9] by mechanical testing, to give directional values of the stiffness coefficients and the technical elastic constants. It should be noted that both experimental methods present intrinsic difficulties;

Table 2
TECHNICAL ELASTIC CONSTANTS OF BONE[32]

	Bovine femur	Human femur
E_1	11.6	13.0
E_2	14.6	14.4
E_3	21.9	21.5
G_{12}	5.29	4.74
G_{13}	6.29	5.85
G_{23}	6.00	6.56
ν^{12}	0.30	0.37
ν^{13}	0.11	0.24
ν^{23}	0.21	0.22
ν^{21}	0.38	0.42
ν^{31}	0.21	0.40
ν^{32}	0.31	0.33

Note: E, G values in GNm^{-2}; ν is dimensionless; 1, radial direction; 2, circumferential direction; 3, longitudinal direction.

the ultrasound method does not provide a direct result, as it involves measurement of the velocity of a propagating wave (with some attendant assumptions[29,30]), while in mechanical testing, it is difficult to obtain adequately-sized specimens away from the longitudinal direction. More recently, both approaches have been extended by Knets and Malmeisters[31] (mechanical testing) and Van Buskirk and Ashman[32] (ultrasound), to a more general and realistic assumption, namely that the symmetry of bone is orthotropic, with the consequent requirement to measure *nine* independent components of the stiffness matrix. These investigations gave similar trends, but significantly different absolute values. As the values for the directional stiffness coefficients and technical constraints of Van Buskirk and Ashman[32] agree more closely with the various studies of the limiting technical constants (e.g. of E in the longitudinal and transverse direction) already discussed, it is suggested that these data, collected in Table 2 for bovine and human compact bone, provide the most reasonable available description of the technical elastic constants of bone.

One difficulty remaining is in deciding the particular strain rate appropriate to the ultrasonic tests — and it would appear to be relatively high; but the E values are in fact similar to those obtained at quasi-static rates. However, it is anticipated that the current high level of research activity on this topic will continue to produce considerable refinements both in the ultrasonic measuring technique and the controlling assumptions.

V. MODELS OF ELASTIC DEFORMATION

During the past two decades, there have been a number of contributions to an interpretation of the Young's modulus of bone in terms of its major microstructural constituents, namely hydroxyapatite and collagen. The original qualitative concept that bone could be considered as a composite of hydroxyapatite-reinforced collagen was suggested by Currey.[33] Bonfield and Li[22,23] subsequently evaluated a composite model for Young's modulus based on the linear rule of mixtures, given by

$$E_b = E_h V_h + E_c V_c \qquad (3)$$

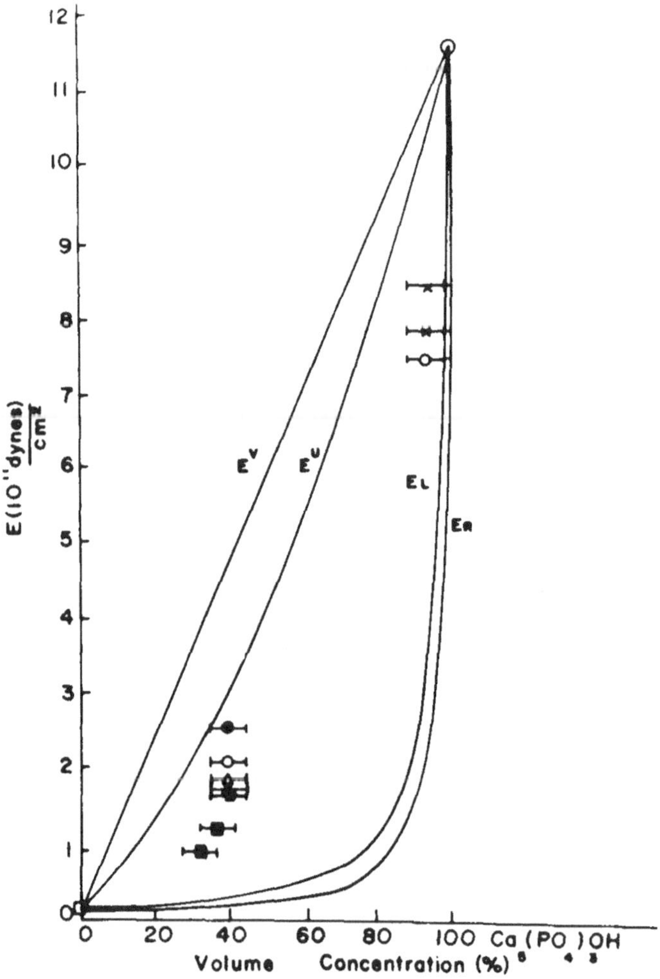

FIGURE 11. Voigt, E^V, and Reuss, E^R bounds on Young's modulus for a hydroxyapatite-reinforced collagen composite (the E^U and E_L bounds can be neglected for the present discussion). The experimental data are (from the left) for collagen, bone, dentine (only slightly higher hydroxyapatite content than that of bone), enamel and hydroxyapatite. (From Katz, J. L., *Mechanical Properties of Bone*, AMD, Vol. 45, American Society of Mechanical Engineers, New York, 1981. With permission.)

where E_b, E_h and E_c are the Young's moduli of bone, hydroxyapatite and collagen, and V_h and V_c are the volume fractions of hydroxyapatite and collagen, respectively. This simple model gave reasonable agreement between the calculated modulus values and the experimental data then available ($E_h \sim 63.5$ GNm^{-2}, $E_c \sim 1.3$ GNm^{-2}). However, Grenoble et al.[29] subsequently measured the modulus of synthetic hydroxyapatite ultrasonically and obtained a significantly higher value ($E_h \sim 114$ GNm^{-2}), which produces a factor of 2 difference between experiment and the predictions of Equation 3. As discussed by Katz,[34] Equation 3, based on the Voigt uniform strain model, in fact provides an upper bound, while the Reuss uniform stress model provides a lower bound, with neither approach satisfying the experimental results, as illustrated in Figure 11. This figure also demonstrates the difficulty in critically testing any particular model by solely examining the dependence of Young's modulus on the volume fraction of hydroxyapatite. As the volume fraction of hydroxyapatite in mature bone only varies within narrow limits, as discussed in an earlier section, then to achieve a sufficiently large variation, it is necessary to include values for materials such as dentine and enamel, an exercise which introduces other structural variations.

Similarly, to include "young" bone would introduce the additional complication of a different level of vascular space (and in practice the hydroxyapatite volume fracture is not very different, e.g. 0.45 in 1-month rabbit tibia compared with 0.50 in 12-month rabbit[13]). This impasse was resolved by a classic paper by Currey[35] in which, from Cox's equation for a fiber-reinforced composite, he proposed a composite model for bone based, not only on V_h and V_c, but also on the hydroxyapatite crystal aspect ratio (length/diameter) and, in particular, on hydroxyapatite orientation, as follows:

$$E_b = E_h V_h \,(1 \quad \tan h\, \beta L/2)/(\beta L/2) + E_c V_c \tag{4}$$

with

$$\beta = ((G_c/E_h)\,(2\pi/A_h \log_e [R/r_o]))^{1/2} \tag{5}$$

where r_o, A_h, L, 2R are the hydroxyapatite crystal radius, cross-sectional area, length, and axis to axis separation, respectively, and G_c is the collagen shear modulus.

Further,

$$E_{b_0}/E_{b_\phi} = \cos^4\phi + (E_{b_0}/E_{b_{90}})\sin^4\phi + [(E_{b_0}/G_b - 2\nu_b)]\cos^2\phi \sin^2\phi \tag{6}$$

where $E_{b\phi}$, E_{b_0} and E_{b90} are the Young's moduli of bone with the hydroxyapatite crystals at angles of $\phi°$, 0° and 90°, respectively, to the stress axis, G_b is the shear modulus and ν is Poissons ratio. From these equations, it is possible to predict the theoretical dependence of E on ϕ based on reasonable assumptions as the morphology of hydroxyapatite, with a result such as shown in Figure 2. The difficulty with this approach, which was readily apparent at the time (when the experimental data of E against ϕ in Figure 2 was not available) is that, due to the large disparity between E_h and E_c, the modulus of bone specimen with $\phi \rightarrow 0°$ is largely determined by E_h while the modulus as $\phi \rightarrow 90°$ is largely controlled by E_c. Hence the model predicts E_{90}/E_o as ~ 0.1, whereas as discussed previously, from experiments, $E_{90}/E_o \sim 0.5$. The later experimental data on E for a range of ϕ values obtained by Bonfield and Grynpas[8] confirmed this difference between experiment and theory, and also permitted further development of a composite model for bone.

From these data, Katz[36] developed the important concept that bone could be modeled as two-level composite, with one level provided by hydroxyapatite-reinforced collagen in a single osteon (Figure 12) and the second level by the approximately hexagonal packing of osteons (i.e. oriented bone) in a matrix of interstitial (i.e. randomly oriented) bone (Figure 13). The physical reality of this model may be judged by comparison with typical micrographs[37] of bovine cortical bone shown in Figures 14 and 15. At the first level, a single osteon is assumed to a coherent entity, with an elastic modulus (E_o) computed on the basis of the single fiber reinforced models (as modified by Katz) discussed earlier, with E_o simply replacing E[illegible] and yielding a similar orientation dependence. The second level provides the important difference in treatment, as E_{bone} is now developed by assuming a near-hexagonal packing of the osteon units, (the reinforcing phase), with the cement line and interstitial bone acting as the matrix phase. Katz then utilized the Haskin-Rosen model[36] for a hexagonally or randomly arranged, *hollow* fiber composite to compute E values for bone, an approach which requires the incorporation of various geometrical factors and mechanical properties of the matrix and fibers, some known and some assumed (e.g. for the cement line). Hence, the absolute value of E_{bone} derived depends on the particular values used in the model, but the important result is that the predicted upper and lower bounds for the ratio

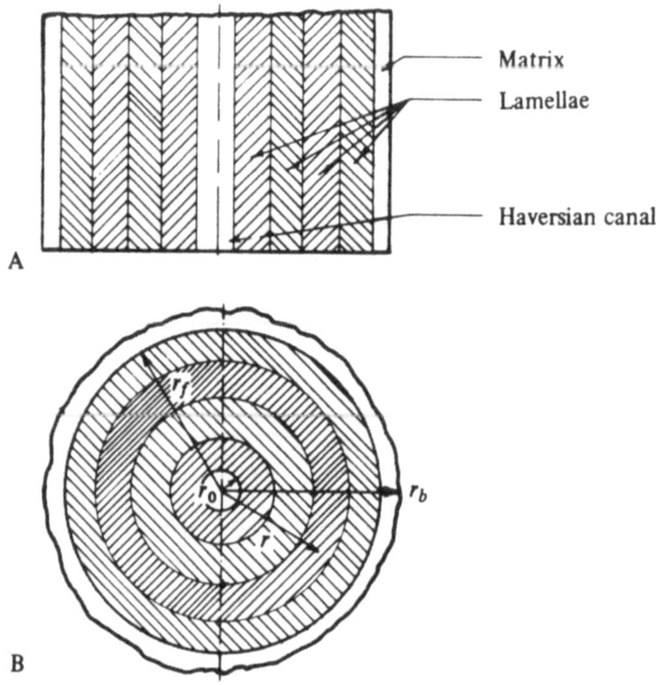

FIGURE 12. A single osteon modeled as a hollow fiber, with each lamella composed of hydroxyapatite-reinforced collagen. (From Katz, J. L., *Mechanical Properties of Bone*, AMD, Vol. 45, American Society of Mechanical Engineers, New York, 1981. With permission.)

FIGURE 13. The near hexagonal packing of osteons in an interstitial bone matrix. (From Katz, J. L., *Mechanical Properties of Bone*, AMD, Vol. 45, American Society of Mechanical Engineers, New York, 1981. With permission.)

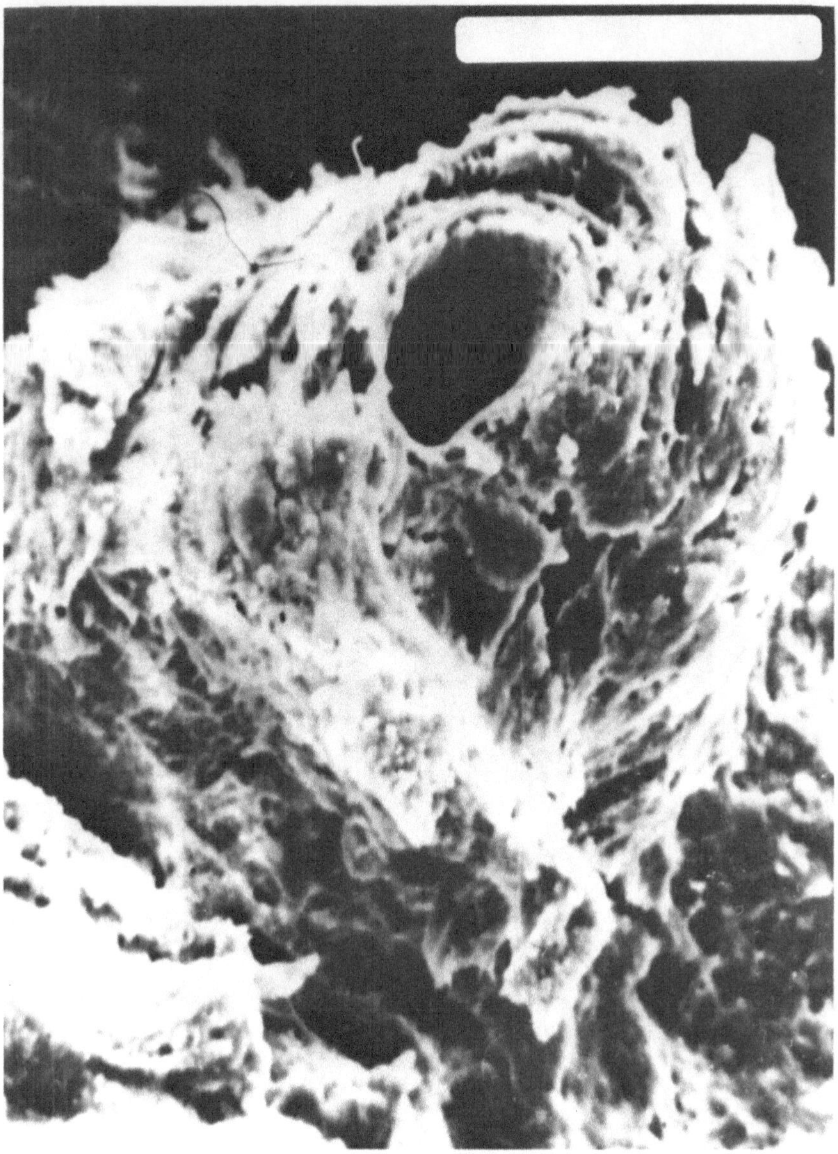

FIGURE 14. A single osteon in a bovine tibia section (marker = 33 μm). (From Behiri, J. C. and Bonfield, W., *J. Mater. Sci.*, 15, 1841, 1980. With permission.)

E_{90}/E_{o} (~ 0.8 to 0.4) are reasonably similar to the experimental value. A comparison of the theoretical predictions for particular assumptions with the experimental data is shown in Figure 16. It should be emphasized that at this stage neither the model nor the experimental results account in detail for the *variability* in bone microstructure, but, such refinements will undoubtedly be made with a continuing research. For a complete description of the deformation of bone, this chapter has demonstrated that the incorporation of a viscoelastic component into the model represents a desirable next step in understanding.

ACKNOWLEDGMENTS

The continuing support of the Science and Engineering Research Council for the program

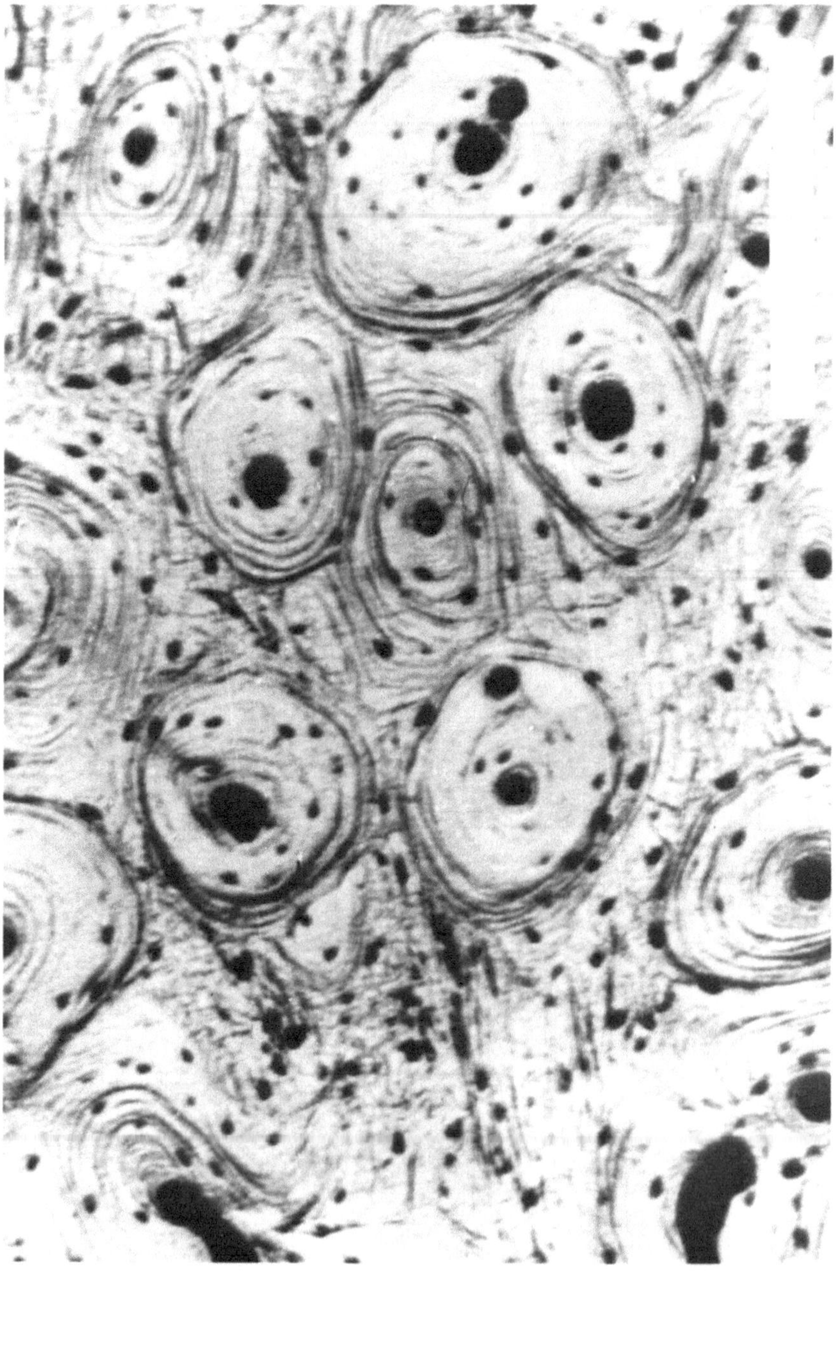

FIGURE 15. Osteon grouping in a section perpendicular to the long axis in bovine tibia cortical bone (marker = 244 μm). (From Behiri, J. C. and Bonfield, W., *J. Mater. Sci.*, 15, 1841, 1980. With permission.)

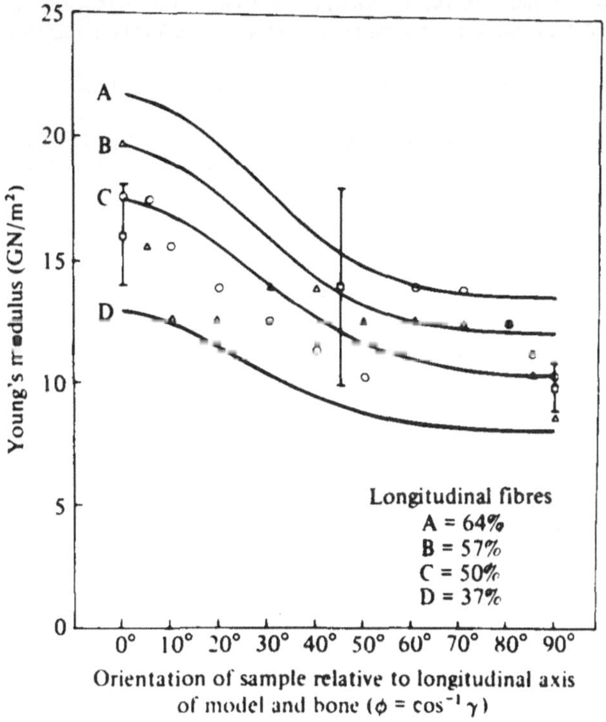

FIGURE 16. Comparison of predictions of the Katz two-level composite model with the experimental data of Bonfield and Grynpas. Each curve represents a different lamellar configuration within a single osteon, with longitudinal fibers: A, 64%; B, 57%; C, 50%; D, 37%; and the rest of the fibers assumed horizontal. (From Katz, J. L. *Mechanical Properties of Bone*, AMD, Vol. 45, American Society of Mechanical Engineers, New York, 1981. With permission.)

of research at Queen Mary College on the deformation and fracture of bone is gratefully acknowledged, together with the contributions over the years of my research students and assistants, Drs. P. O'Connor, P. K. Datta, M. D. Grynpas, A. Nash, J. C. Behiri, A. E. Tully, and E. A. Clark.

REFERENCES

1. **Evans, F. G.,** *Mechanical Properties of Bone*, Charles C Thomas, Springfield, Ill., 1973.
2. **Katz, J. L.,** The structure and biomechanics of bone, in *The Mechanical Properties of Biological Materials*, Cambridge University Press, 1980.
3. **Bonfield, W. and O'Connor, P.,** Anelastic deformation and the friction stress of bone, *J. Mater. Sci.*, 13, 202, 1978.
4. **Burstein, A. H., Currey, J. D., Frankel, V. H., and Reilly, D. T.,** The ultimate properties of bone tissue: the effects of yielding, *J. Biomech.*, 5, 34, 1972.
5. **Bonfield, W.,** Mechanisms of fracture in bone, in *Mechanical Properties of Bone*, AMD, Vol. 45, American Society of Mechanical Engineers, New York, 1981.
6. **Dempster, W. T. and Liddicoat, R. T.,** Compact bone as a non-isotropic material, *Am. J. Anat.*, 91(3), 331, 1952.
7. **Bonfield, W. and Datta, P. K.,** Young's modulus of bone, *J. Biomech.*, 9, 131, 1976.

8. **Bonfield, W. and Grynpas, M. D.,** Anisotropy of the Young's modulus of bone, *Nature,* 270, 453, 1977.
9. **Reilly, D. T. and Burstein, A. H.,** The elastic and ultimate properties of compact bone tissue, *J. Biomech.,* 8, 393, 1975.
10. **Bonfield, W. and Tully, A. E.,** Ultrasonic analysis of the Young's modulus of cortical bone, *J. Biomed. Eng.,* 4, 23, 1982.
11. **McElhaney, J. H.,** Dynamic response of bone and muscle tissue, *J. Appl. Physiol.,* 21, 1231, 1966.
12. **Crowinshield, R. D. and Pope, M. H.,** The response of compact bone in tension at various strain rates, *Ann. Biomed. Eng.,* 2, 217, 1974.
13. **Bonfield, W. and Clark, E. A.,** Elastic deformation of compact bone, *J. Mater. Sci.,* 8, 1590, 1973.
14. **Currey, J. D.,** The effects of strain rate, reconstruction and mineral content on some mechanical properties of bovine bone, *J. Biomech.,* 8, 81, 1975.
15. **Wright, T. M. and Hayes, W. C.,** Tensile testing of bone over a wide range of strain rates: effects of strain rate microstructure and density, *Med. Biol. Eng.,* 14, 671, 1976.
16. **Bonfield, W. and Datta, P. K.,** Impact fracture of compact bone in a shock tube, *J. Mater. Sci.,* 9, 1609, 1974.
17. **Torzilli, P. A., Burstein, A. H., Takebe, K., Zika, J. C., and Heiple, K. G.,** The material and structural properties of maturing bone, in *Mechanical Properties of Bone,* AMD, Vol. 45, American Society of Mechanical Engineers, New York, 1981.
18. **Currey, J. D.,** The mechanical consequences of a variation in the mineral content of bone, *J. Biomech.,* 2, 1, 1969.
19. **Abendschein, W. G. and Hyatt, G. W.,** Ultrasonic and selected physical properties of bone, *Clin. Orthop.,* 69, 294, 1970.
20. **Ascenzi, A. and Bonucci, E.,** Mechanical similarities between alternate osteons and cross-ply laminates, *J. Biomech.,* 9, 65, 1976.
21. **Bonfield, W. and Li, C. H.,** Deformation and fracture of bone, *J. Appl. Phys.,* 37, 869, 1966.
22. **Bonfield, W. and Li, C. H.,** Anisotropy of non-elastic flow in bone, *J. Appl. Phys.,* 38, 2450, 1967.
23. **Bonfield, W. and Li, C. H.,** Temperature dependence of the deformation of bone, *J. Biomech.,* 1, 323, 1968.
24. **Lanyon, L. E.,** Analysis of surface bone strain in the calcaneus of sheep during normal locomotion, *J. Biomech.,* 6, 41, 1973.
25. **Carter, D. R., Harris, W. H., Vasu, R., and Caler, W. E.,** The mechanical and biological response of cortical bone to in vivo strain histories, in *Mechanical Properties of Bone,* AMD, Vol. 45, American Society of Mechanical Engineers, New York, 1981.
26. **Currey, J. D.,** Anelasticity in bone and echinoderm skeletons, *J. Exp. Biol.,* 43, 279, 1965.
27. **O'Connor, P.,** Deformation and fracture behaviour of compact bone, Ph.D. Thesis, University of London, 1976.
28. **Lang, S. B.,** Ultrasonic method for measuring elastic coefficients of bone and results on fresh and dried bovine bones, IEEE Trans., *Biomed. Eng.,* 17, 101, 1970.
29. **Grenoble, D. E., Katz, J. L., Dunn, K. L., Gilmore, R. S., and Murty, K. L.,** The elastic properties of hard tissues and apatites, *J. Biomed. Mater. Res.,* 6, 221, 1972.
30. **Yoon, H. S. and Katz, J. L.,** Ultrasonic wave propagation in human cortical bone — I. Theoretical considerations for hexagonal symmetry, *J. Biomech.,* 9, 407, 1976.
31. **Knets, I. V. and Malmeisters, A.,** The deformability and strength of human compact tissue, in *Mechanics of Biological Solids,* Bulgarian Academy of Sciences, Sofia, 1975.
32. **Van Buskirk, W. C. and Ashman, R. B.,** The elastic moduli of bone, in *Mechanical Properties of Bone,* AMD, Vol. 45, American Society of Mechanical Engineers, New York, 1981.
33. **Currey, J. D.,** Three analogies to explain the mechanical properties of bone, *Biorheology,* 2, 1, 1964.
34. **Katz, J. L.,** Hard tissue as a composite material I. Bounds on the elastic behaviour, *J. Biomech.,* 4, 455, 1971.
35. **Currey, J. D.,** The relationship between the stiffness and mineral content of bone, *J. Biomech.,* 2, 477, 1969.
36. **Katz, J. L.,** Composite material models for cortical bone, in *Mechanical Properties of Bone,* AMD, Vol. 45, American Society of Mechanical Engineers, New York, 1981.
37. **Behiri, J. C. and Bonfield, W.,** Crack velocity dependence of longitudinal fracture in bone, *J. Mater. Sci.,* 15, 1841, 1980.

Chapter 5

VISCOELASTIC PROPERTIES OF BONE

Roderic S. Lakes and J. Lawrence Katz

TABLE OF CONTENTS

I. INTRODUCTION

Viscoelastic response in a material may be taken to include all recoverable effects in which the local stress at an instant of time depends not only on the strain at that instant, but on the *history* of the strain. History-dependent effects resulting in permanent changes in the material are analyzed within the framework of plasticity and fracture mechanics; these effects in bone are not discussed in the present chapter. Viscoelastic effects occur in all biological materials studied to date, with the possible exceptions of dental enamel and echinoderm skeletons. Viscoelastic behavior in bone has been the subject of many investigations during the past century. There is a large variety of motivation for this kind of study. For example, the static load-carrying performance of structural members is influenced by their viscoelastic response, if any. Since bone supports load, it is of interest to know how this function is altered by viscoelasticity. Furthermore, bone in the body is subjected to loads which vary with time. Impulsive loads are damped out to a degree related to the viscoelastic behavior of the bone in question. Since impulsive loads are thought to damage articular cartilage, a knowledge of the damping qualities of different types of normal and diseased bone is useful. By inquiring into the causes of bone's viscoelastic behavior, it is possible to learn what happens to the mechanical energy which is dissipated in the deformation of bone in the body and over what time scale. A variety of mechanisms has been suggested to account for the Wolff's law activity by which bone remodels itself in response to stress so as to support the stress more effectively. These mechanisms entail direct and/or indirect coupling of deformation energy of bone to the cells responsible for remodeling so that an appropriate stimulus is provided. Therefore, investigation of viscoelastic mechanisms can be useful in understanding the causes of Wolff's law related activity. Recently, increasing use of orthopedic implants has raised questions regarding the compatibility between time-dependent mechanical properties of the implant and those of the surrounding tissues. It is therefore necessary to know the viscoelastic properties of both bone and implant to answer this kind of question most effectively. In particular, a constitutive equation for bone is necessary in the construction of mathematical models of bone-implant systems. Finally, a knowledge of the viscoelastic properties of normal and diseased bone, even at frequencies which are not present in the normal use of bone, is useful to experimenters who are developing diagnostic tools based on wave propagation or mechanical vibrations in bone. Such techniques are potentially useful for assessment of the mechanical quality of bone in vivo and are intended to provide information unavailable through standard radiographic techniques.

In this chapter, the results of different viscoelastic measurements are compared in light of the environmental and experimental parameters now known to affect such measurements.

II. BACKGROUND: VISCOELASTICITY IN SOLIDS

A. The Viscoelastic Functions and Their Meaning

The history-dependent character of the response to stress of materials such as bone, may be described in many ways. These descriptions are not independent; the mathematical interrelationships among them are discussed in Ferry[1] and Gross.[2] The principal viscoelastic functions used in the present study are as follows:

1. Relaxation modulus, G(t)*: If a specimen of a viscoelastic material such as bone, is subjected to a strain which is applied suddenly and held constant, the stress will decrease

* The symbol G has been traditionally used to denote a shear modulus. Stiffnesses associated with other stress states, such as the Young's modulus E for tension/compression, are dealt with in a similar fashion [E(t), E′(ω), E*(ω)]. E and G do not in general depend on time (frequency) in the same way.

with time t. The modulus G(t) is the ratio of stress to (constant) strain and represents a stiffness which decreases with time.

2. Creep compliance, J(t): If the specimen is subjected to a stress which is applied suddenly and held constant, the strain will increase with time. The creep compliance J(t) is the ratio of strain to (constant) stress, and also generally increases with time.

3. Dynamic moduli, $G'(\omega)$, $G''(\omega)$, $G^{\star}(\omega)$: When a viscoelastic solid is subjected to an oscillatory strain of frequency $\omega/2\pi$, where ω is the angular frequency, the ratio of stress to strain will have an in-phase portion G' and an out of phase portion G''. The overall stiffness, $G^{\star}$ = square root of $G'^2 + G''^2$, depends on frequency. G', the storage modulus, is related to the mechanical energy stored in oscillatory loading, and G', the loss modulus, is related to the mechanical energy which is dissipated.

4. Loss tangent, tan δ: This is defined by $\tan\delta = G''/G'$ and therefore also depends on frequency. The ratio of energy dissipated in a quarter cycle of dynamic loading, to the energy stored, is just $(\pi/2)\tan\delta$. The attenuation of stress-waves in a material is proportional to tan δ. Specifically, the wave attenuation α, is given by $\alpha = (\tan\delta/2)\,\omega/c$ where c is the speed of the wave. The wave speed c is given, for longitudinal waves in a rod, by c = square root of $E^{\star}/\rho$ sec $\delta/2$, where ρ is the material density,[3] Therefore, the "shock absorbing" capability of bone is directly related to the value of its loss tangent.

B. The Boltzmann Superposition Integral

If a material is subjected to a step-function strain history (in one dimension), $\epsilon(t) - \epsilon_0 H(t)$, where H(t) is the unit step function and t is time, the resulting stress history can be written $\sigma(t) = G(t) \cdot \epsilon_0$, where G(t) is a function characteristic of the material and is called the relaxation modulus. Boltzmann's Superposition Principle states that the effect of a compound cause is the linear superposition of the effects of the individual causes. When a "box function", strain history $\epsilon(t) = \epsilon_0[H(t-t_1) - H(t-t_2)]$, is considered to be the compound cause,* then the resulting stress history must be:

$$\sigma(t) = [G(t - t_1) - G(t - t_2)]\epsilon_0, \text{ for } t > t \tag{1}$$

Now, consider an arbitrary strain history $\epsilon(\tau)$ which is zero before time $\tau = 0$, then the stress at time t due to an element of strain history between time $t - \tau$ and $t - \tau + d\tau$ can be expressed as:

$$d\sigma(t) = [G(t - \tau + d\tau) - G(t - \tau)]\epsilon(\tau) = \frac{dG(t - \tau)}{d\tau}\,\epsilon(\tau)\,d\tau \tag{2}$$

The total stress now at time t is the limit of the sum (integral) of all such expressions as d becomes arbitrarily small, i.e.,

$$\sigma(t) = \int_0^t \frac{dG(t - \tau)}{d\tau}\,\epsilon(\tau)d\tau + \epsilon(t)\,G(0) \tag{3}$$

where the second term is the stress due to the strain at the present time, which does not relax. Integrating this by parts yields the Boltzmann Superposition integral in its usual form, a constitutive equation for a linearly viscoelastic material:

* This corresponds to a relaxation and recovery experiment, as shown in Figure 1.

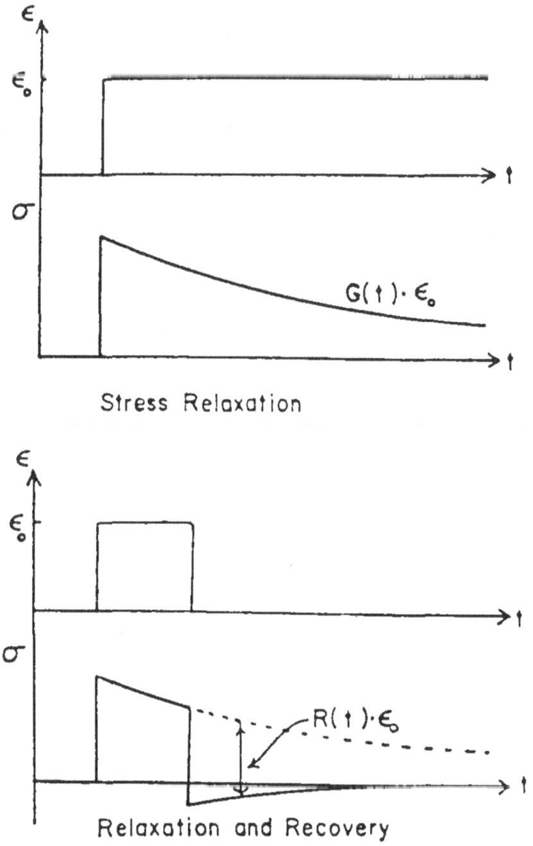

FIGURE 1. Strains and stresses associated with stress relaxation and with relaxation and recovery. G(t) is the relaxation modulus and R(t) is the recovery modulus. For linearly viscoelastic materials they are equal.

$$\sigma(t) = \int_0^t G(t - \tau) \frac{d\epsilon(\tau)}{d\tau} d\tau \quad (4)$$

$$\sigma_{ij}(t) = \int_0^t C_{ijkl} (t - \tau) \frac{d\epsilon_{kl}(\tau)}{d\tau} d\tau \text{[three-dimensional form, anisotropic solid]} \quad (5)$$

This last expression is the simplest one capable of adequately describing the viscoelastic behavior of bone. Its significance lies in the fact that once the kernel function C_{ijkl} (or G for shear in one dimension) is known, the stress response to *any* strain history, however complex, can be calculated. The kernel can be obtained by experiments employing *simple* strain histories (e.g., sine wave, step function). The results of such different types of experiments can be converted into a common representation by means of relations readily derivable from Equation 4. In addition, linearly viscoelastic materials are characterized by stiffness parameters which depend on time (or frequency) but not on strain level. This simplicity of behavior is not found in materials which are nonlinearly viscoelastic.

Equation 4 has a number of special cases which for reasons of conceptual simplicity have appeared frequently in the literature of bone mechanics. One of the simplest of these special cases is to assume that $G(t) = ae^{-t/\tau} + b$, which corresponds to a single relaxation time process. Such behavior is frequently associated with the spring-dashpot model depicted in Figure 2. In this type of model, the elastic or energy storage aspects of the material are

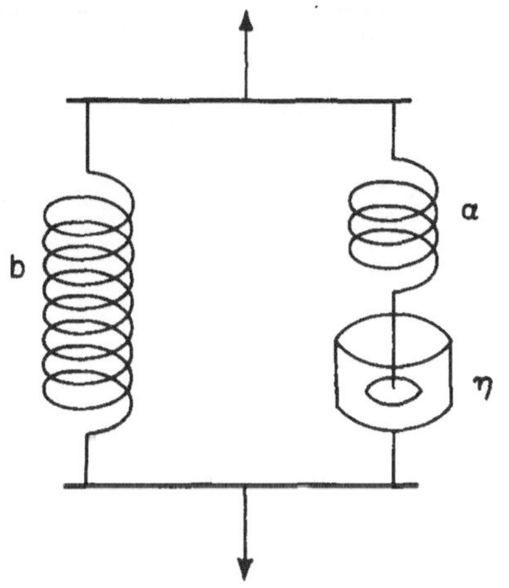

FIGURE 2. A three-element spring-dashpot model for simple viscoelastic behavior. The single relaxation time is $\tau = \eta/a$.

lumped in the spring elements, while the viscous or energy dissipative aspects are lumped in the dashpot element. As shown in Figure 3, the dynamic functions associated with this single relaxation time model (for $a = 0.25$, $b = 1$, $\tau = 1$) vary with frequency only over a limited portion of the spectrum. Since the description of real materials (including bone) using spring-dashpot models, generally requires an infinite number of such elements, a formulation employing a continuous spectrum of relaxation times is more appropriate.

The spectrum of relaxation times $H(\tau)$ is related to the measurable linear viscoelastic functions by the following relationships:

$$G(t) = Ge + \int_{-\infty}^{\infty} H(\tau)e^{-t/\tau} d\log\tau \quad (6)$$

$$G'(\omega) = Ge + \int_{-\infty}^{\infty} H(\tau)\frac{\omega^2\tau^2}{1+\omega^2\tau^2} d\log\tau \quad (7)$$

$$G''(\omega) = \int_{-\infty}^{\infty} H(\tau)\frac{\omega\tau}{1+\omega^2\tau^2} d\log\tau \quad (8)$$

where $G(t)$ is the relaxation modulus, $G'(\omega)$ is the dynamic storage modulus, $G''(\omega)$ is the dynamic loss modulus and $Ge = \lim_{t\to\infty} G(t)$. These integrals may be inverted with some difficulty to obtain the spectrum from the measured quantities; however, in practice, approximations are generally used. The role of the spectrum in the structure of viscoelasticity theory is treated by Gross.[2]

For the case in which the spectrum is arbitrarily sharp [i.e., a delta function, $H\tau/\tau = \Delta G\delta(\tau - \tau_0)$] the following are obtained (the so-called Debye equations):

$$G(t) = Ge + \Delta Ge^{-t/\tau_0} \quad (9)$$

$$G''(\omega) = Ge + \Delta G\frac{\omega\tau_0}{1+\omega^2\tau_0^2} \quad (10)$$

$$G''(\omega) = \Delta G\frac{\omega\tau_0}{1+\omega^2\tau_0^2} \quad (11)$$

Such expressions are obtained theoretically for situations in which there is a single characteristic rate at which the system in question readjusts itself to equilibrium, i.e., a single

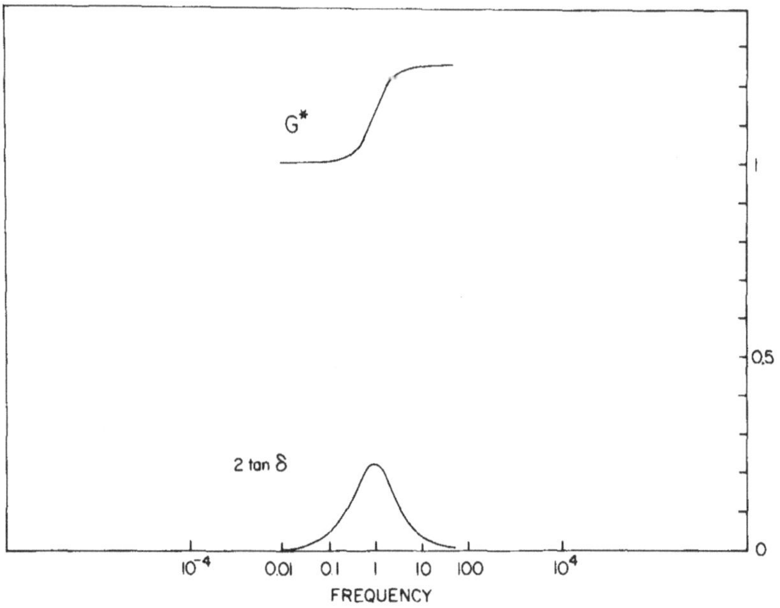

FIGURE 3. The dynamic functions G★ and tan δ for a single relaxation time process.

relaxation time. Experimentally measured quantities generally have a more gradual time or frequency dependence than is predicted by the Debye equations.

The relaxation spectrum is useful in the development of constitutive laws and will be discussed in more detail later. Conceptually, the spectrum's significance is that it provides a measure of "relaxation activity" or the "strength" of viscoelastic response. In a domain of time scale where the spectrum is small, the material in question behaves quasi-elastically.

III. EXPERIMENTAL STUDIES OF BONE VISCOELASTICITY

Interest in the behavior of bone as a structural element in the human body has motivated a large number of investigations into the mechanical properties of bone; for extensive and comprehensive reviews, the interested reader is referred to the reviews by Evans,[4] Katz and Mow,[5] and Reilly and Burstein.[6] Compact bone has proven to be a highly complex material. It is inhomogeneous, anisotropic,[7-9] viscoelastic,[10] and exhibits detailed structure on all levels of scale.[11]

Investigations which attempt to go farther than a simple demonstration of viscoelastic effects in bone have not been done until relatively recently. The following discussion of these investigations will for the most part be confined to experiments done on wet bone, since it is known that drying significantly affects the properties of bone.[12]

A. Uniaxial Tension/Compression Parallel to the Bone Axis, Compact Bone

Smith and Keiper[13] examined the dynamic behavior of wet specimens of human long bones, over the frequency range 500 to 3500 Hz at a strain of 10^{-4}; the temperature range is not given. The storage modulus E′ had an average value of 15.9 GN/m²* and was independent of frequency within the experimental resolution. The loss tangent obtained from the measured loss modulus averaged .025 for four femoral bone specimens. Thompson[14] determined the dynamic modulus and loss angle of strips of wet canine radial bone by

* 1 GN/m² is equivalent to 0.145×10^6 lb/in².

examining the specimen's behavior in transverse resonant vibration. The dynamic modulus averaged 15 GN/m^2, and the loss tangent was ≅ 0.016 in the frequency range 168 to 3500 Hz. Both functions were essentially frequency independent, within experimental scatter. Neither temperature nor strain-level were quoted. Bargren, Bassett, and Gjelsvik[15] determined the dynamic moduli of wet human femoral bone at three frequencies: 0.66, 5.2, and 7.5 Hz; the modulus E^* was ~ 16 GN/m^2 and varied little with frequency; the loss tangent was quite small, ~ .003. Temperature and strain-level were not stated. Laird and Kingsbury[16] measured the dynamic moduli of wet bovine humeral bone at room temperature. The storage and loss moduli both exhibited a rather jagged frequency dependence in the range 1 to 16 kHz. Peak acceleration of the vibrator in the testing machine, was kept constant. Since the acceleration is the second time derivative of the displacement, the strain amplitude in the specimen varied by a factor $16^2 = 256$ over the frequency range studied. The acceleration amplitude was not quoted; therefore the possibility of nonlinear behavior or even yielding in the specimens, cannot be excluded. Black and Korostoff[17] in a dynamic study of wet, viable, human tibial bone at 37°C (body temperature), also reported a jagged frequency dependence of the modulus. Anomalous behavior, i.e., a sudden change of the modulus with frequency, near 200 Hz was seen in some specimens. Now, since the preceding results have been obtained directly by dynamic techniques, the dynamic modulus $|E^*|$ and the dynamic loss tan δ can be compared straightforwardly.

In a series of experiments primarily aimed at examining the fracture properties of bone, Currey[18] writes the following constitutive equation for the Young's modulus E of wet bovine femoral bone in tension as a function of strain-rate ϵ: $E = 22.8 - 5.47\,H + 1.01 \log_{10}(10^4\epsilon)$. H is the fraction of Haversian tissue present and the units for E are GN/m^2. Over a three-decade range in strain rate, 1.3×10^{-4} to 0.16/sec, the modulus was found to change by 14%. Since one sample was used for each datum, the data exhibited considerable scatter. A linear regression analysis was therefore used to obtain the above equation. Earlier, McElhaney[19] performed compressional constant strain-rate experiments on wet bovine bone and expressed his results as a family of stress vs. strain curves at six different strain rates between 10^{-3} and 1500/sec. The initial portion of each of these curves is a straight line. These data have been transformed into the complex modulus representation by Lakes and Katz.[20] The results exhibit considerably larger loss tangents and dispersions (variation in modulus with frequency) than those reported elsewhere. The "bumps" in these curves are artifacts resulting from applying the transformation technique to such a small ensemble of data points. Recently, Bonfield and Datta[21] subjected a specimen of wet bovine tibial bone to compressional loading at room temperature. At a strain-rate of 3×10^{-4}/sec, the stress strain curve in contrast with those obtained by McElhaney[19] is concave downward. The "initial" slope of this curve, i.e., the tangent modulus is for small strains, 27.3 GN/m^2, while the secant modulus corresponding to a strain of 2×10^{-3} is 20% lower. Bonfield and Datta[21] state that their results imply nonlinear behavior. Tennyson et al.[22] have examined the compressive properties of wet bovine femoral bone at strain rates between 10 and 450/sec using a split Hopkinson bar apparatus. They modeled their data using the equation $\sigma = E\epsilon + \eta\, d\epsilon/dt$, where η is a viscosity parameter which corresponds to the Voigt viscoelastic element. For experiments performed just following the death of the animal, $E = 3 \times 10^6$ lb/in^2[20.7 GN/m^2], and $\eta = 8$ $lb\text{-}sec/in^2$ [5.5×10^4 $Nsec/m^2$] were obtained; however, the strain level was not quoted. 43 samples were used for 83 tests. Since the specimens were subjected to rather large stresses (40,000 lb/in^2 or 276 MN/m^2) the use of some samples for more than one experiment is questionable. Yielding occurs in bone compression at stresses between 18,000 and 40,000 lb/in^2 (124 to 276 MN/m^2) depending on strain rate.[19]

Lugassy and Korostoff[23] have determined the stress-relaxation properties of wet bovine femoral bone in compression; however, no sample temperature was quoted. Direct data obtained at a strain of 5.1×10^{-4} were taken from Lugassy[24] and transformed into the

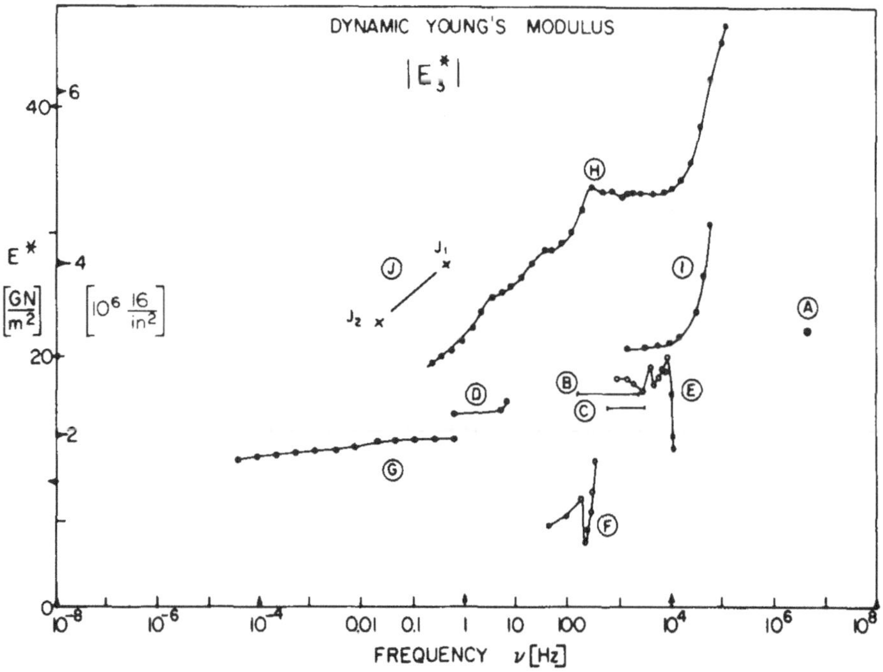

FIGURE 4. Comparison of dynamic Young's moduli for bone, measured along the long axis of the bone. (A) From Lang.[7] Wet bovine phalanx. Temperature not stated. Ultrasonic. (B) From Thompson.[14] Wet canine radius. Transverse resonant vibration. (C) From Smith and Keiper.[13] Wet human bone. Dynamic. (D) From Bargren, Bassett, and Gjelsvik.[15] Wet human femur. Dynamic. (E) From Laird and Kingsbury.[16] Wet bovine humerus. Room temperature. Dynamic, using accelerometers. (F) From Black and Korostoff.[17] Wet, viable, human tibia. Body temperature (37.5°C). Dynamic. (G) From Lugassy.[24] Wet bovine femur. Calculated from relaxation data by Lakes and Katz.[20] (H) From Mc-Elhaney.[19] Bovine. Calculated from constant strain-rate data by Lakes and Katz.[20] (I) From Tennyson, et al.[22] Wet bovine femur. No temperature given. (J) From Bonfield and Datta.[21] Wet bovine tibia. No temperature given. Calculated from constant strain-rate data. J_1: Tangent modulus, assuming $\epsilon = 10^{-4}$. J_2: Secant modulus, assuming $\epsilon = 2 \cdot 10^{-3}$.

dynamic modulus representation by Lakes and Katz.[20] Lugassy tested his apparatus with steel samples which do not relax significantly and found that relaxation in the apparatus itself introduced some errors at short times following the application of the strain (corresponding to the higher frequencies in the transformed data). At the very high end of the frequency spectrum, Lang[7] used ultrasonic techniques to obtain the elements of the "elastic" modulus tensor and the Young's moduli in different directions, for specimens of wet bovine phalanx; no temperature was quoted. Similar studies on human cortical bone have been used to study the anisotropy on the microstructural level of organization by Yoon and Katz.[8,9] Ultrasonic properties of cortical bone are not discussed in this chapter.

The results of many of these measurements of the compressive or tensile Young's modulus and loss tangent for bone have been plotted in Figure 4 and 5, respectively. Note the considerable lack of agreement among these results.

Anisotropy in compression of human cortical bone was also observed in quasi-static experiments by Dempster and Liddicoat.[25] These data, as well as measurements of breaking strengths, were compared with corresponding values for common woods, stones, and metal. The stiffness of bone exceeds that of most woods, and its strength exceeds that of most woods and stones, as well as brick and concrete. Young's modulus tests were done at a stress rate of 82.8 MN/m^2 sec; "wet" samples were obtained by soaking previously dry material for 24 hr.

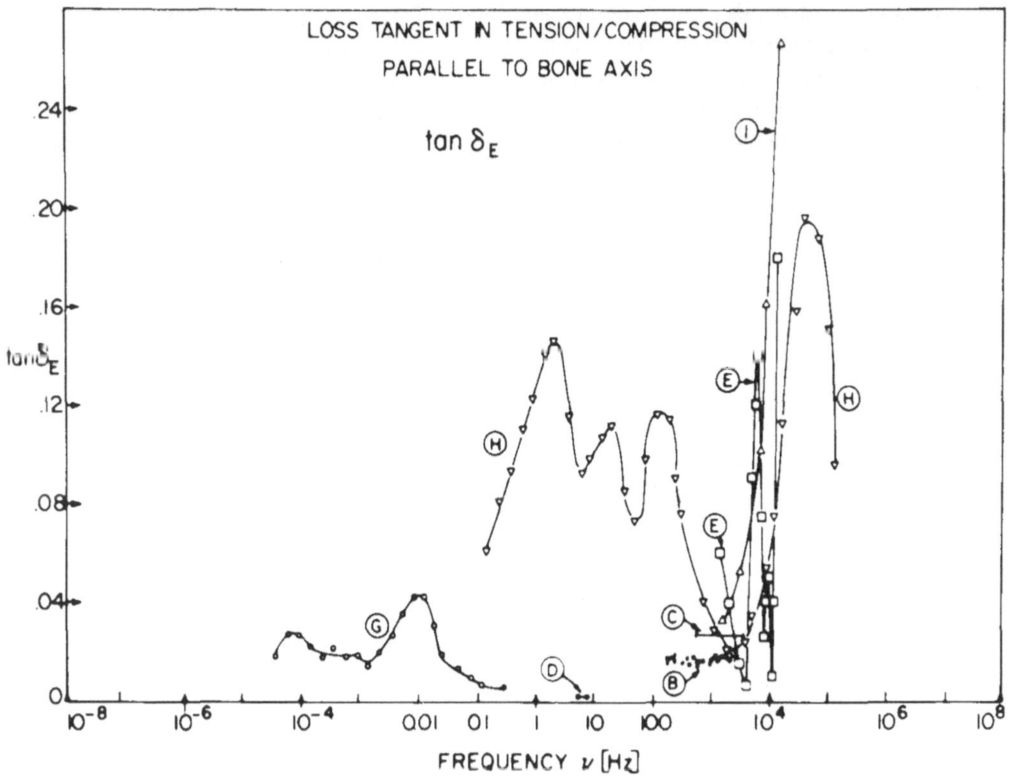

FIGURE 5. Comparison of dynamic loss tangents in tension/compression for cortical bone. Letter designations correspond to those in Figure 4. (A) None measured. (B) From Thompson.[14] Wet canine radius. Transverse resonant vibrator. (C) From Smith and Keiper.[13] Wet human bone. Dynamic. (D) From Bargren, Bassett, and Gjelsvik.[15] Wet human femur. Dynamic. (E) From Laird and Kingsbury.[16] Wet bovine humerus. Room temperature. Dynamic, using accelerometers. (F) None inferred from data in Figure 4. (G) From Lugassy.[24] Wet bovine femur. Calculated from relaxation data by Lakes and Katz.[20] (H) From McElhaney.[19] Bovine. Calculated from constant strain-rate data by Lakes and Katz.[20] (I) From Tennyson et al.[22] Wet bovine tibia, no temperature given. Calculated from constant strain rate data. (J) None inferred from data in Fig. 4.

Some variation in the magnitude of the modulus is expected as a result of variations in degree of calcification, in porosity, as well as in species. For example: bovine bone is found to be stiffer than human bone provided that identical experimental procedures are used to measure both. (See, e.g., Reilly and Burstein.[6]) The large variation in the frequency dependence of the modulus and of the loss tangent are less readily explained. The shape, but not the overall frequency dependence, of these two modulus curves, is sensitive to this kind of error. The "bumps" in the curves obtained from McElhaney's data,[19] in particular, must be considered to be artifacts; however, the steep rise of the transformed modulus with frequency and the large loss tangent, are unaffected by transformation artifacts. Nonlinear viscoelastic behavior, if present in bone, can also cause this kind of disagreement. This can come about in two ways. First, the transformation procedure depends on the assumption of linear response. Secondly, moduli measured directly by dynamic methods can be expected to differ in magnitude and frequency dependence if the measurements are performed at different strain levels upon a nonlinear material. The question of nonlinear behavior in bone is discussed in more detail in a subsequent section of this chapter.

Jaggedness in the graphs of the dynamic modulus vs. frequency obtained by Black and Korostoff[17] and by Laird and Kingsbury,[16] is of interest because such behavior cannot arise from relaxation effects describable by linear viscoelasticity theory. In linear solids, the most abrupt variation of the dynamic modulus and loss with frequency is that associated with a single relaxation time phenomenon. A jagged modulus vs. frequency curve can result from nonlinear resonant behavior in the specimen itself or from resonances in the testing apparatus. Since machine resonances are likely to occur in the frequency range considered here, great care must be taken to ensure that such resonances do not obtrude in the data. For example, it is advisable to examine a specimen having known mechanical properties using the testing apparatus in question. Since the distribution of machine resonances will depend on the geometric stiffness (force divided by displacement) of the standard specimen, this stiffness should be comparable to that of the unknown material.

B. Bending

Currey[26] examined the long-term creep behavior of wet bovine tibial and metacarpal bone specimens in cantilever bending and made the following observations:

1. Creep, measured in the range 2 min to 10 days and plotted vs. log time, gives a curve which is always concave up (room temperature).
2. Recovery following this test appeared to be asymptotic toward zero strain after 100 days.
3. In repeating the creep experiment, the maximum difference in the deflections was only 5% initially and almost nothing later. After 55 days, the creep curve was still concave up through the entire time range.
4. Dry bone exhibited less creep than wet, but irreversible effects due to drying were not marked.
5. Creep deflection at a given time was proportional to load.
6. Creep deflection after 2 min and 2 hr depends on temperature in the range 2.5 to 48°C.
7. At body temperature, recovery following one day of creep is asymptotic to zero strain; in the range 2 min to 54 days, the creep curve (deflection vs. log time) is always concave upward (Figure 6). Deflection at the end of 54 days is 24 times the "initial" deflection. Attack by bacteria and fungi was prevented by a crystal of thymol dissolved in the water bath.

C. Shear, Compact Bone

The earliest measurements of shear moduli for bone were done by Rauber,[10] who did not note time-dependent behavior. Bonfield and Li[27] observed shear moduli in the range 0.81 to 0.88×10^6 lb/in^2 (5.6 to 6.1 GN/m^2) for filaments of bovine tibial bone in torsion about the bone axis. The temperature and degree of hydration of the specimens were not given. The rate of angular deformation was 1 degree/min. From this, the strain rate was calculated to be 5.8×10^{-6}/sec, given the specimen length and diameter 0.25 inch and 0.010 inch, respectively. For a strain of 10^{-3}, the effective frequency corresponds to $\nu_{eff} = \dot{\epsilon}/\epsilon \cdot 2\pi = 0.93 \times 10^{-3}$H. Lang[7] determined the shear modulus of wet bovine phalanx at high frequency (3MHz) using an ultrasonic technique; Yoon and Katz[8,9] determined it for human bone at 2 MHz. Adler and Cook[28] found the ultrasonic attenuation of wet canine tibial bone at 22°C. The polarization of the ultrasonic waves was not given, so that it is uncertain whether the measurement gives $\tan \delta_{2323}$ or $\tan \delta_{1313}$. Thompson[14] directly measured the shear loss tangent of whole, dry canine radii at several audiofrequencies, using a resonant torsional vibration technique. Although dry bones were used, the results are included here since they are the only ones available in the mid-audiofrequency range.

Finally, Lakes, Katz, and Sternstein[29,30] performed dynamic and stress-relaxation measurements on wet specimens of human and bovine compact bone at a variety of temperatures. A typical set of data for human tibial bone at body temperature and at small strains appropriate for linear behavior was selected. A constitutive equation was then obtained to describe the

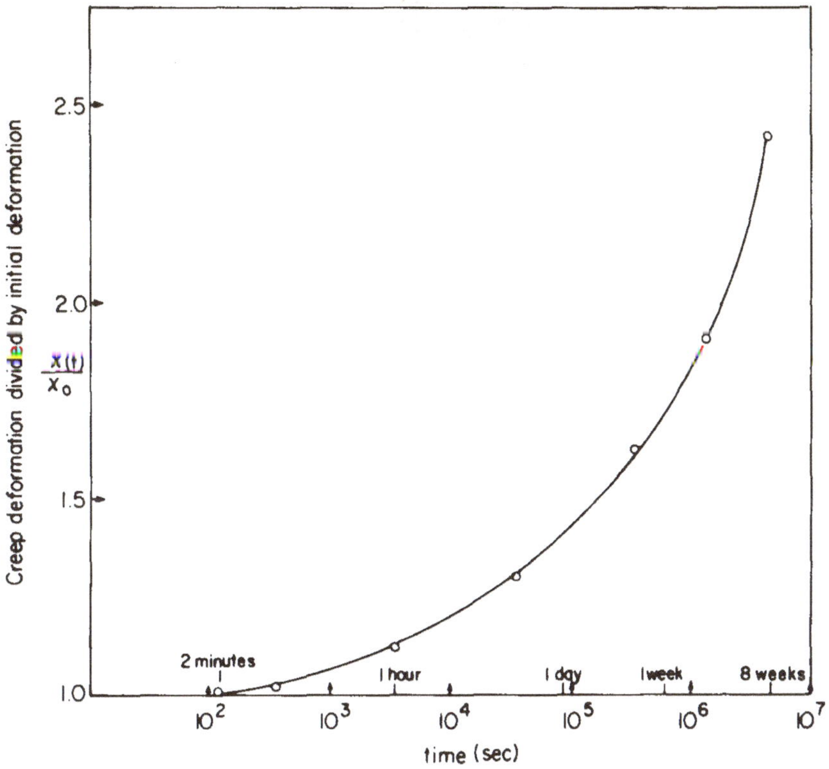

FIGURE 6. Bending creep in bone. Replotted from Currey.[18]

dynamic and relaxation data,[31] and dynamic moduli were calculated analytically from this equation in the low frequency domain below 0.5 Hz. These results, as well as results from the other shear experiments described above, are plotted in Figure 7. Agreement among these results appears to be better than for the compressional results; however, there were fewer sets of data to compare in the shear case, so this may not be significant.

D. The Effect of Temperature

Bone in the body performs its functions at or near body temperature (37°C). Nevertheless, experiments are sometimes performed on bone at other temperatures for reasons of simplicity in the laboratory set-up, or to learn something about the causes of various effects observed in bone. Since much of the available experimental information has been obtained at temperatures other than body temperature, it is necessary to evaluate the effects of temperature changes on the mechanical properties of bone. The temperature range over which bone can be examined in an unaltered state is limited from above by the denaturation temperature of collagen which, when shielded in bone may be about 100°C; and from below by the temperature at which the tissue fluids freeze, somewhere below 0°C.

Smith and Walmsley[32] observed that the Young's modulus of compact bone decreases linearly by approximately 0.51% per degree C increase in temperature. The effects of temperature changes on the "elastic" properties of bone were found to be completely reversible in the range 9°C to 43°C (40°F to 110°F). Sedlin[12] found that the modulus of elasticity of wet compact human bone measured in cantilever bending at a slow strain rate, decreased by 6% as temperature was increased from 21°C to 37°C, corresponding to a change of 0.38° in stiffness per degree C. Currey[18] performed cantilever bending creep tests on wet bovine compact bone. He observed a similar change of 0.4% per degree C, or 8%, between 17° and 37°C in the "immediate" deflection following the application of a constant load.

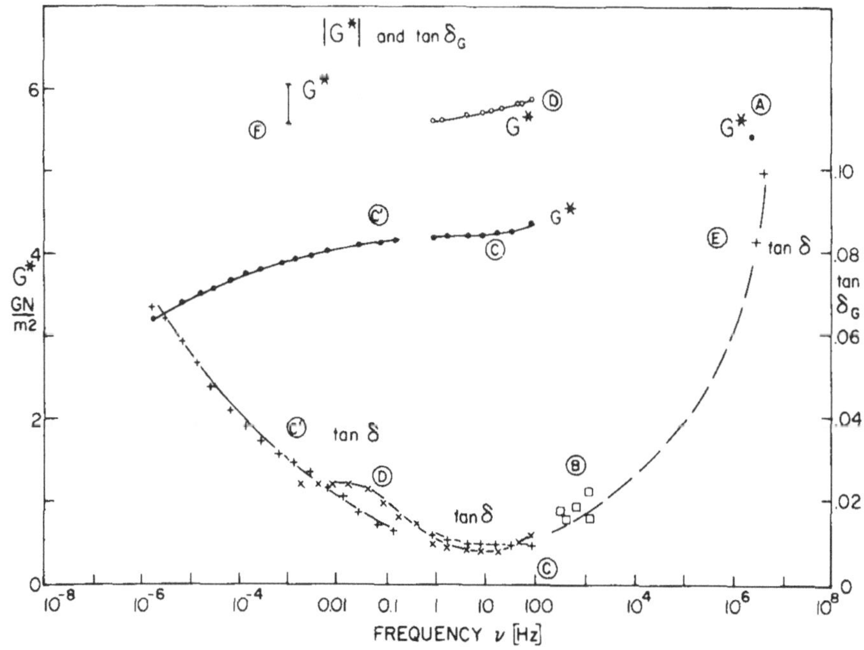

FIGURE 7. Comparison of shear properties of cortical bone. (A) From Lang.[7] Wet bovine phalanx. Temperature not stated. Ultrasonic. (B) From Thompson.[14] Dry canine radius. Resonant vibration. (C) From Lakes, Katz, and Sternstein.[29] Wet human tibia. Body temperature (37°C). Driven, dynamic. (C') From Lakes, Katz, and Sternstein.[29] Wet human tibia. Body temperature (37°C). Calculated from relaxation data using a constitutive equation developed by Lakes.[73] (D) From Lakes, Katz, and Sternstein.[29] Wet bovine femur. Body temperature (37°C). Driven, dynamic. (E) From Adler and Cook.[28] Wet canine tibia, 22°C. Calculated from ultrasonic attenuation. (F) From Bonfield and Li.[27] Bovine tibia filament. Temperature and degree of hydration not stated. Shear modulus at constant strain rate. Frequency calculated from given strain rate, for $\epsilon = 10^{-3}$.

In contrast, the additional creep deformation following 2 hr at constant load increased by 100% over the same temperature range. Moreover, the creep deformation did not increase linearly with temperature; rather it increased more rapidly at higher temperatures. In a series of torsional dynamic and stress relaxation studies on wet bovine and human bone, Lakes, Katz, and Sternstein[29,30] observed that the stress-relaxation behavior at long times is more significantly affected by temperature changes than dynamic or short-term relaxation behavior. In addition, the temperature-dependent effects observed were thermorheologically complex; i.e., a change in temperature does not simply shift the graph of viscoelastic functions along the log-time or log-frequency axis — the graph is also distorted. The results of a room-temperature viscoelastic experiment on bone are therefore not related in a simple fashion to the behavior observed at body temperature.

Temperature changes appear to have a relatively minor effect on "elastic" properties of bone measured at relatively low frequencies or at short times. The effect on viscoelastic response, particularly over long periods of time, is significant. However, little information is available regarding the effect of temperature on the viscoelastic behavior of bone at relatively high frequencies (above 100 Hz).

E. The Effect of Viability

As pointed out in other portions of this treatise, bone is a living, dynamic tissue. Never theless, the fraction of the volume of compact bone occupied by living cells is relatively small. For example the lacunae in which the osteocytes reside, represent 0.8%[33] of the total volume, and the Haversian canals, which contain nerves or blood vessels, account for 4 to 6% of the total volume of compact bone.[34] Furthermore, the cellular components are mechanically more compliant by many orders of magnitude than the extracellular matrix. In view of these considerations, most, but not all, investigators have been content to use dead bone for measurements of mechanical properties.

Tennyson et al.[22] have examined the dependence of the elastic and viscous mechanical behavior of wet bovine femoral bone, on the time following the death of the animal. They described their results using a three-element model and noted that the elastic modulus of bone attained a final value 33% lower than the initial value and that the "viscous" modulus decreased by 67% after a period of 16 days post-mortem. The specimens used in these experiments had been stored in plain water rather than Ringer's solution, therefore irreversible changes in the extracellular matrix due to dissolution or recrystallization of the mineral phase, cannot be ruled out. Furthermore, the use of same specimens for more than one test may have resulted in irreversible changes due to yielding as discussed earlier. Black and Korostoff[17] obtained specimens of human tibial bone from amputated limbs and subjected these specimens to dynamic mechanical testing under physiological conditions. The viability of several specimens was demonstrated by means of tissue culture. Re-testing of the specimens following cell death for the purpose of comparison was, however, not done. However, Black[35] has reported subsequently on the differences between "live" and dead bone. Recently, Fitzgerald[36] studied the dynamic mechanical behavior of trabecular bone and of muscle as a function of post-mortem time. Specimens were immersed in Ringer's solution to provide a near-physiologic environment. Changes of 12.3% in the storage compliance and 26.2% in the loss compliance were seen after 24 hr "dead time" in specimens of human cancellous bone. In dead bone, the effects of prolonged soaking in Ringer's solution appear to have a negligible effect on the viscoelastic properties for immersion times of one week or less (Lakes, Katz, and Sternstein[30]) and a minor but measurable effect for times up to one month.[37]

A dependence of mechanical properties on viability or post-mortem time is not surprising in the case of trabecular bone, for which a substantial portion of the volume is occupied by soft tissue. A similar dependence in the case of compact bone would be less readily explained, except perhaps for appropriate stress states with nonzero dilatation, in a range of frequency where the viscoelastic behavior is significantly affected by "fluid-flow" processes. Unfortunately, the experimental results available to date regarding compact bone cannot be regarded as conclusive.

F. Nonlinear Behavior

As discussed earlier, the linear theory of viscoelasticity is based on the postulate that the effect of a compound cause is the superposition of the effects of the individual causes considered alone. The phenomena of yielding and fracture are, for example, manifestly nonlinear. Compact bone in "normal" use functions at stress levels well below the yield stress, therefore the question of possible nonlinear behavior under such circumstances is not obvious and must be dealt with experimentally. Experiments designed to disclose nonlinear behavior in a material are of necessity more complex than those intended for known linear materials. For example, a material obeying a constitutive equation of nonlinear superposition (in one dimension):

$$\sigma = \int_0^t G(t - \tau, \epsilon) \frac{d\epsilon}{d\tau} d\tau \tag{12}$$

must be subjected to tests at many strain levels in order to extract the kernel G which characterizes the behavior of the material. Some nonlinear materials behave in a yet more complex fashion. For example, the general nonlinear viscoelastic theory of Green and Rivlin[38] can be expressed in one dimension for small strains as the following:[39]

$$\sigma(t) = \int_0^t G_1(t - \tau) \frac{d\epsilon}{d\tau} d\tau + \int_0^t \int_0^t G_2(t - \tau_1, t - \tau_2) \frac{d\epsilon}{d\tau_1} \frac{d\epsilon}{d\tau_2} d\tau_1 d\tau_2 + \text{——} \quad (13)$$

To characterize such a material it is necessary not only to perform experiments at different strain levels but also to examine the response of the material to complex strain histories. For example, in the time domain, histories consisting of superpositions of step-functions have been used for characterizing polymers (e.g., see Pipkin and Rogers[40]). The simplest such history is a "box function" strain, which corresponds to a relaxation and recovery experiment. In the frequency domain, an appropriate generalization of the usual dynamic experiments entails the application of a strain history containing a sum of oscillations at different frequencies. For materials describable by Equation 12, the relaxation and recovery moduli are equal at a given strain; the stress response to the multi-frequency history contains integer multiples of each frequency in the strain program.[41] By contrast, materials obeying Equation 13 recover at a different rate from that at which they relax, and multifrequency strain histories generate interactions (sum and difference terms) between frequencies.[42]

In most experimental studies of compact bone, linear viscoelastic response has been assumed rather than demonstrated. Therefore, experimental procedures were often used which, although adequate to describe known linear materials, were quite insensitive to the presence of possible nonlinear effects. For linear materials, the results of different types of viscoelastic experiments can be mathematically converted into a common representation for purposes of comparison and for obtaining a description of the response of the materials over a wider domain in time or frequency than would be possible using a single type of experiment. Lakes and Katz[20] performed calculations of this type and found substantial disagreement among different measurements of compressional viscoelastic response in compact bone. Furthermore, several sets of published data were found to be internally inconsistent with linear viscoelasticity theory. From those observations, the above authors concluded that compact bone in compression appears to behave nonlinearly, even at "small" strains.

Bonfield and Datta[21] performed compressional experiments on wet compact bovine bone at a single constant strain-rate. They observed that the stress-strain curve obtained in these experiments was concave downward and that the secant modulus at a strain 2×10^{-3} was 20% smaller than the tangent Young's modulus for small strains less than 2.3×10^{-4}. They concluded that this curvature of the stress-strain graph was a result of nonlinear behavior in bone. It is appropriate to point out here that this conclusion would be valid only if bone were perfectly elastic. However, bone exhibits viscoelastic behavior so that the curvature in the graph can result either from nonlinearity, or from relaxation effects, or from both. Indeed, for linearly viscoelastic materials, the slope of the stress-strain curve is simply the relaxation modulus (e.g., see Ferry[1]):

$$[d\sigma_{33}/d\epsilon_{33}]_t = E_3(t) \quad (14)$$

Since E(t) is generally monotonically decreasing, the stress-strain graph for a linearly viscoelastic material will be concave downward, so that an inference of nonlinear behavior from a single curved graph of this type, is unwarranted.

Investigations of the torsional viscoelastic behavior of compact bone, with attention to various types of nonlinear effects, were performed by Lakes, Katz, and Sternstein.[29,30] For human bone, the nonlinearities were predominantly "strain-dependent", i.e., describable by an equation of nonlinear superposition.[31] The shear relaxation modulus increased by ~3% as the strain was increased from 8.5×10^{-4} to 1.7×10^{-3}. In human bone this "shear stiffening'' was relatively insensitive to time in relaxation (the relaxation curves were nearly parallel). However, this was not true for bovine bone. Time-dependent nonlinearity consisting of recovery occurring more slowly than relaxation, was also observed. This effect was relatively small and was most noticeable at long times; it was accentuated by the presence of a superposed axial tensile stress in biaxial experiments.

IV. CAUSES FOR BONE VISCOELASTICITY

When a viscoelastic solid is subjected to mechanical strain, part of the strain-energy is converted to other forms and is ultimately degraded into heat. What happens to this energy in living bones can be of considerable biological importance. For example, bone is known to remodel itself in response to mechanical stress in such a way as to more effectively support the stress (Wolff's law). A variety of mechanisms have been proposed to account for this activity:

1. Lamellar motions, impingement on osteocyte processes.[43]
2. Stress-induced fluid motion, resulting in improved nutrition of osteocytes.[44,45]
3. Stimulation of bone-cells by means of stress generated potentials.[46]

Currently, the third of these is favorably accepted by many investigators as the origin of Wolff's law activity. In all the above processes, mechanical energy is transformed into other forms before being dissipated. An investigation of these and other viscoelastic mechanisms, in particular of their frequency dependence, can elucidate the role of strain energy dissipation in living tissues.

Among the large number of physical processes which are capable of causing viscoelastic* effects, the following are considered in this section:

1. Inhomogeneous deformation (motion of osteons at cement lines, motion of lamellae in osteons and interstitial lamellae, and motion of fibers within a lamella).
2. Molecular modes in collagen.
3. Thermoelastic coupling (homogeneous effect and inhomogeneous effect).
4. Piezoelectric-like coupling.
5. Motion of fluid in canals in bone.

The mechanical damping which results from the last three of these processes can be calculated theoretically. An outline of the basses for these calculations is presented below.

A. Relaxation Due to Thermoelasticity

1. Homogeneous Effect

The stiffness measured for a solid is dependent upon whether the measurement is performed under adiabatic or isothermal conditions. If the observation is done at frequencies at which the solid behaves neither adiabatically nor isothermally, a frequency dependence of the stiffness will be observed, as well as mechanical damping resulting from the irreversible flow of heat between the specimen and its (isothermal) environment. The adiabatic and isothermal compliance tensors S^{S}_{ijkl} and S^{T}_{ijkl}, respectively, are related by the following:[47]

* Viscoelastic effects are here taken to include any process which results in energy dissipation in mechanical deformation at small strains, and which does not permanently alter the structure of the material, regardless of cause. Several of the effects considered here depend on specimen size and are not included in the traditional study of viscoelasticity theory; nevertheless, they result in macroscopically observable damping and hence must be included in the analysis of bone behavior.

Table 1
SOME RELEVANT PHYSICAL PROPERTIES OF BONE

Substance	Property	Value	Units	Conditions	Ref.
Human bone	Volume fraction		%		33
	canaliculi	1.48	%		
	lacunae	.80	%		
Human femur	Cross-section area	6.1	%		34
Human fibula	of spaces	4.5			
Bone	Thermal expansion	2×10^{-5}	°C	Parallel to bone axis	48
		3×10^{-5}	°C	Perpendicular to bone axis	
Human tibia	Thermal conductivity	7.1×10^{-4}	Joule/°K cm sec	Wet	49
Hydroxyapatite	Heat capacity, c_p	770.52	Joule/mole°K	T = 298.16°K	50
Bone	Piezoelectric coefficients				
	d_{123}	.130	pcoul/Nt	Dry	51
	d_{213}	−.203	pcoul/Nt		
Bone	Dielectric constant K'_{11}	5.9	—	Dry bone,	52
	Dielectric loss K''_{11}	0.1	—	1 kHz to 100 kHz	
Human bone	Compliance S_{2323}	11.48×10^{-11}	M^2/Nt	Dry, ultrasonic	9

$$S^S_{ijkl} - S^T_{ijkl} = -a_{ij}\, a_{kl} \frac{T}{C^\sigma} \tag{15}$$

where a is the thermal expansion tensor, T the absolute temperature, and C^σ the heat capacity per unit volume at constant stress. In a principal coordinate system in which a is diagonal, the adiabatic and isothermal shear compliances (such as S_{2323}) are identical, so that the homogeneous thermoelastic effect causes no relaxation or damping in shear.

For an experiment done in uniaxial tension or compression, one which measures S_{3333}, for example, the compliances differ, and the frequencies at which the maximum energy loss occurs depends on the geometry:

$$\tan \sigma_{ijkl}(\omega) = \Delta_{ijkl}\, \Sigma_n F_n \frac{\omega\, \tau_n(ijkl)}{1 + \omega^2 \tau_n^2(ijkl)} \text{ with } \Sigma_n F_n = 1 \tag{16}$$

and

$$\Delta_{ijkl} = (S^S_{ijkl} - S^T_{ijkl})/S^T_{ijkl}$$

and F_n and τ_n are solutions of an eigenvalue problem for the geometry in question.[53] For a circular cylinder of radius r in compression along the long axis, Zener has obtained $F_n = 4/q_n^2$, $\tau_n = r^2/q_n^2\, v/k$ where C_v is the specific heat, k is the thermal conductivity, and q_n the n'th zero of the zeroth Bessel Function J_0. For a bone specimen 1/8 inch in diameter at body temperature, using the values for S, σ, and k compiled in Table 1 and the specific heat for hydroxyapatite in the absence of a value for bone, the loss tangent due to the first mode has a maximum value of 0.0014 at a frequency of 0.013 Hz, corresponding to a relaxation time of 12 sec. Losses due to the first three modes are plotted vs. frequency in Figure 8. The characteristic times τ_n depend on the specimen's size and their distribution upon the specimen's shape; therefore the relaxation should not be thought of as ''intrinsic''. Nevertheless, the calculated loss tangents entail a real loss of mechanical energy.

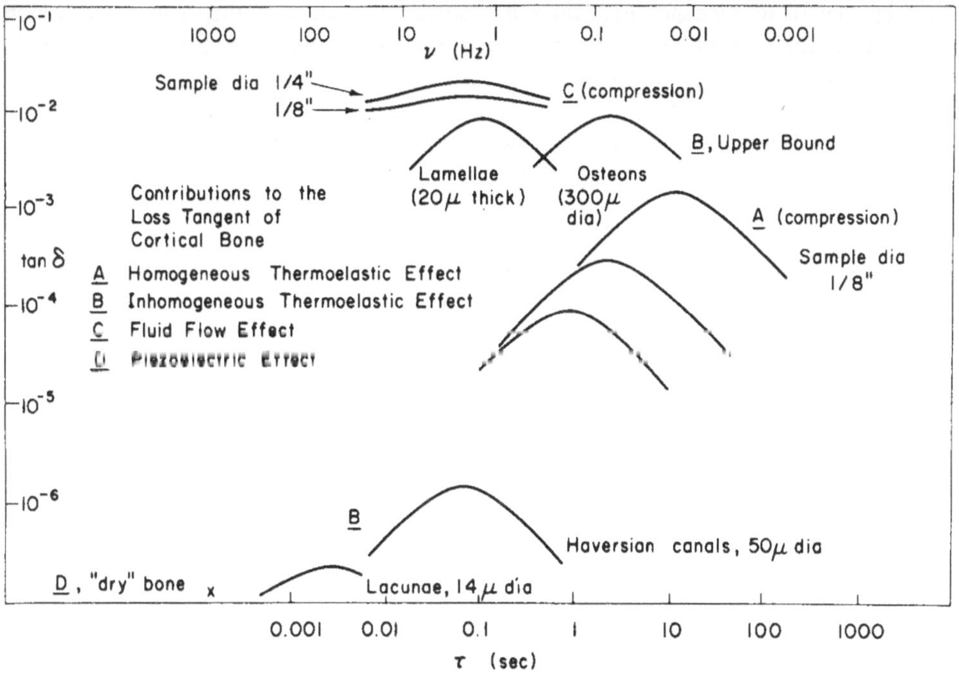

FIGURE 8. Predicted contributions to the mechanical losses in cortical bone.

2. Inhomogeneous Effect

Stress inhomogeneities in a solid which is subjected to dynamic loading, give rise to fluctuations in temperature and therefore to heat flow between the inhomogeneities. This heat flow entails an irreversible conversion of mechanical energy into heat; this energy dissipation is represented in terms of the loss tangent as follows:[53]

$$\tan \delta = \frac{\beta^2 T}{C_v X} R \sum_k F_k \frac{\nu_k \nu}{\nu_k^2 + \nu^2} ; \sum_k F_k = 1 \tag{17}$$

where β is the volume coefficient of thermal expansion, T the (absolute) temperature, C_v the heat capacity at constant volume, per unit volume; and R is the fraction of strain-energy associated with dilatation. The characteristic frequencies ν_k and weighting coefficients F_k are obtained from analysis of the specific geometry in question.

The quantity $\Delta = \beta^2 T/C_v X$ represents an upper bound for the loss in a given material, since $R \leq 1$. For compact bone, this expression was evaluated by Lakes and Katz[41] using the data in Table 1. The compressibility may be expressed in terms of the elastic constants by $X = S_{1111} + S_{2222} + S_{3333} + 2(S_{1122} S_{2233} + S_{1133})$. The heat capacity used in the calculation is again that of hydroxyapatite in the absence of a value for a whole bone. For bone at body temperature, then, $\Delta = 8.1 \times 10^{-3}$.

In order to evaluate the quantity R, a stress analysis must be performed on the inhomogeneities in question. This apparently cannot be done for bone, since detailed information regarding the anisotropy and boundary constraints of the different types of osteons and lamellae is not available. Nevertheless, note that Ascenzi and Bonucci[54] have observed differences as great as a factor of two in the Young's modulus of different types of osteons, apparently depending upon differences in the respective microstructures. This greatly exceeds the variation in stiffness of the crystallites of lead due to random orientation. Therefore, the value of R for bone is likely to be significantly greater than that for lead (0.065), but it could not be expected to attain the upper bound of 1 corresponding to pure local dilatation.

The characteristic frequency ν_0 of these thermoelastic losses, i.e., the frequency at which tan δ is maximum, depends only on the thermal relaxation time for heat flow between the inhomogeneities (of size L) in question. ν_0 is given by $\nu_0 = k/L^2C_v$ where k is the thermal conductivity, C_v the heat capacity, and L is the thermal path length. Predicted thermoelastic losses resulting from variations in the stiffness of osteons and lamellae are plotted in Figure 8 on the basis of an upper bound value of R = 1.

The presence of cavities in a solid also gives rise to a relaxation of thermoelastic origin, as a result of inhomogeneous stress distributions around the cavities. Losses of this type calculated by Lakes and Katz[41] on the basis of the Zener relaxation theory are at least four orders of magnitude smaller than losses actually measured in bone; viscoelastic response due to stiffness variations, in contrast, could be large enough to detect.

B. Relaxation Due to Stress-Generated Potentials

The stiffness of an elastic crystal is increased by the presence of piezoelectric coupling. This stiffening effect may be expressed in terms of the difference between the compliance at constant electric displacement, S^D_{ijkl} and the compliance at constant electric field, S^E_{ijkl}:[47]

$$S^D_{ijkl} - S^E_{ijkl} = d_{nij}\, d_{mkl}\, [K_{mn}]^{-1}_{\sigma} \tag{18}$$

where d is the piezoelectric modulus tensor, K is the dielectric tensor at constant stress, and the temperature is assumed to be constant.

If the material in question is semiconducting, or exhibits dielectric relaxation, and is subjected to a transient (step-function) stress, the stress-generated polarization will be neutralized after a period of time. Electrical energy is dissipated, resulting in a relaxation of the mechanical stress.

Mechanical losses induced by stress-generated potentials, as well as the piezoelectric-like stiffening referred to above, may be expected to depend upon the electrical boundary conditions imposed on the specimen. Therefore, Lakes and Katz[41] assumed a typical experimental geometry, that of a thin plate, with the one direction perpendicular to its surface; and derived the following expression for the mechanical loss due to piezoelectric coupling:

$$\Delta\tan\delta^{Mech}_{2323} = \frac{d'^2_{123} - d''^2_{123}}{k'_{11}\, S''_{2323}} \tan\delta^{(k)}_{11} \tag{19}$$

where, for lossy materials, the coefficients d and k are considered to be complex:

$$d^{\star} = d' - id''; \; k^{\star} = k' - ik'' \tag{20}$$

and where $\tan\delta^{(k)}_{11} = k''_{11}/k'_{11}$ is the dielectric loss tangent.

For "dry" bone, the piezoelectric-like coefficients for shear, d_{123}, d_{213}, greatly exceed those for compression. Using the published values for the properties of compact bone, which are collected in Table 1, the piezoelectric-like contribution to the loss becomes $\Delta\tan\delta_{2323} = 5\cdot10^{-8}$. The contribution $\Delta\tan\delta_{1313}$ is somewhat larger: $1.2\cdot10^{-7}$. This is far too small to be resolved above other losses, typically of the order 10^{-2}, in bone. Recently more data have become available regarding the variation of the piezoelectric-like coefficients with frequency and humidity;[55] nevertheless, these coefficients do not appear to be large enough to generate measurable mechanical damping. Relaxation due to an inhomogeneous piezoelectric-like effect, in which stress-related polarizations are generated and annihilated locally, is conceivable, given the structure of bone. No experimental data on such an effect appear to be available.

C. Relaxation Due to Fluid Motion

A sizable fraction of the volume of even compact bone consists of voids. A dissipation of mechanical energy can result from the motion of fluids in connected voids if the strain applied produces a volume change and if the specimen has a free surface. Strain-induced fluid-motion has been suggested as a mechanism for certain elements of the 'piezoelectric-like' tensor of wet bone[56] and for facilitated transport of nutrients to and wastes from, osteocytes.[44,45]

The problem of determining the mechanical losses which result from fluid flow in bone was approached by Mouradian,[44] who unfortunately used the bulk modulus of air rather than that of water to perform the calculations. More recently, Nowinski and Davis[57] employed Biot's theory of consolidation to calculate the stress-relaxation response of bone in bending. They calculated a total relaxation of 14%. With the approximations they used, this behavior was describable by a single relaxation time; however, the predicted value of this characteristic time was not stated. Lakes and Katz[41] calculated the fluid-flow related mechanical loss of a cylindrical specimen of compact bone in tension/compression, using the theory of Rusch[58] and Gent and Rusch.[59] For a given pore size the mechanical loss is approximately but not exactly, describable by a single relaxation time model. The theory predicts a size effect in the fluid-flow related losses; the maximum loss tangent increases with the size of the specimen. Typical curves of calculated loss vs. frequency are plotted in Figure 8 assuming a "background" loss of .01 for the solid matrix. Observe that for suitable specimen geometry (thick specimens) the predicted losses will be quite large, comparable to experimentally observed losses. The characteristic frequency is sensitive to the size of the pores (Haversian canals in this calculation) of the bone and can therefore be expected to differ among different specimens. Fluid flow related losses do not occur in shear,[41,57] however, because there are no local volume changes to cause the motion of fluid.

D. Inhomogeneous Deformation

Microscopic examination of bone fracture surfaces has revealed that slowly moving cracks, in particular, tend to propagate along the cementing lines.[60] This observation has led some authors to postulate that the "cement-line" material, primarily composed of protein-polysaccharides, behaves as a "zone of weakness" and acts to retard some types of fracture and accelerate others. If this material exhibits a large compliance or is in fact viscous at small macroscopic stresses, then motion along these interfaces can result in large mechanical relaxations.

In polycrystalline metals, a similar process, grain-boundary slip, leads to large relaxations of the order 40%, as shown theoretically by Zener[61] and experimentally by Ke.[62] Zener was able to carry out a mathematical analysis of the problem by assuming viscous grain boundaries (which cannot support shear stress in the limit of long times) and grains in the shape of interlocking polyhedra. The complex geometry of the microstructure of bone, including anastomoses of osteons, has thus far precluded any attempt to evaluate theoretically the relaxation strength associated with inhomogeneous deformation. Experimental studies by Tischendorf[43] involved microscopic examination of specimens of human bone subjected to stress in cantilever bending; small motions of the lamellae of a magnitude comparable to the resolution of the optical microscope were reported. However, the time-scale associated with this motion was not specified. Now, the large viscoelastic effects observed in bone in bending[26] and in torsion[30,37] over long periods of time, suggested to the present authors that cement-line motion may dominate the behavior of bone in this region of time-scale. Therefore, Lakes and Saha[63] subjected specimens of wet compact bone to a constant torsional stress at body temperature and observed cement-line displacements of over six microns following prolonged loading. Cement-line motions contribute significantly to long-term creep and relaxation in compact bone; however, a quantitative assessment of the magnitude of this contribution and its relationship to nonlinear response remains to be made.

E. Discussion

Although the question of why bone is viscoelastic has not been answered fully, a variety of relevant physical processes have been examined theoretically and experimentally. Among these, homogeneous thermoelastic coupling and fluid flow effects are active only in deformation modes which involve a volume of change (such as compression) and thereby do not contribute to mechanical losses in shear. Both these effects result in losses which are large enough to measure; the fluid flow mechanism produces predicted losses which are quite large, particularly for large specimens. Inhomogeneous thermoelastic effects could also result in significant losses in either tension/compression or in torsion. Experimental data obtained in torsion suggest that this mechanism does contribute to observed losses. The problem of determining relaxation effects which result from interfacial (e.g., cement line) motion has thus far been too formidable to approach theoretically. Recent experimental results suggest that such processes are important when loads are applied to bones over long periods of time. However, the picture is not yet complete. Piezoelectric-like coupling results in viscoelastic losses which are far too small to measure, in that portion of the frequency domain for which data are available.

The observation that collagen alone exhibits viscoelastic behavior is suggestive of a possible role for molecular motions in the collagen phase of bone in determining the viscoelastic response of bone. Since most investigators of this behavior in collagen have used material (e.g., rat tail tendon), which has a microstructure different from that of bone collagen, the application of such results to the present problem is far from straightforward. Molecular motions which result in viscoelasticity in polar polymers have been investigated by dielectric relaxation experiments. For wet bone, ionic conduction and interfacial polarization appear to dominate the dielectric behavior.[64] Therefore, an alternative approach, such as nuclear magnetic resonance, may be more useful in the experimental investigation of molecular motions in collagen.

V. CONSTITUTIVE MODELING

Prior to a successful theoretical treatment of biomechanical problems, a mathematical description of the relation between stress and strain for the tissue in question must be available. However, many experimenters in the past have not attempted to formulate such descriptions, possibly because their data covered only small segments of the time, frequency, and strain domains. Furthermore, as discussed earlier, experimental data for bone in tension/compression exhibit substantial disagreement, so that a consistent constitutive equation describing these data would be difficult if not impossible to arrive at. Available results obtained in shear show better agreement. Therefore, in this section a constitutive equation for the shear properties of bone is developed.

Various empirical distributions of relaxation times have been compiled by Gross.[2] For example, the "box distribution",

$$\underset{B[\tau_1,\,\tau_2]}{H(\tau)} = \begin{Bmatrix} 1 & \tau_1 \leq \tau \leq \tau_2 \\ 0 & \tau > T_2,\ \tau < \tau_1 \end{Bmatrix} \tag{21}$$

has been found to be useful in the description of polymeric solids[65] and of soft tissues.[66] The measured viscoelastic quantities corresponding to the box spectrum may be expressed in terms of well-known functions:

$$\underset{B[\tau_1,\tau_2]}{G(t)} = Ge + [Ei(-t/\tau_1)] - Ei(-t/\tau_2)] \tag{22}$$

$$\underset{B[\tau_1,\tau_2]}{G'(\omega)} = Ge + \frac{1}{2}\log\left[\frac{1 + \omega^2\tau_2^2}{1 + \omega^2\tau_1^2}\right] \tag{23}$$

$$\underset{B[\tau_1,\tau_2]}{G''(\omega)} = [\tan^{-1}\omega\tau_2 - \tan^{-1}\omega\tau_2] \tag{24}$$

where Ei(x) is the well known and extensively tabulated exponential-integral function (e.g., see Lowan[67]), and the subscript $B[\tau_1,\tau_2]$ denotes the two time parameters in the box spectrum. Now, provided that $\tau_2 \gg \tau_1$, the box distribution gives rise to a loss G″ which is essentially constant in the middle of the domain, while $G'(\omega)$ and G(t) decrease linearly with log $1/\omega$ and log t, respectively. For $\tau_2 \geqslant \tau_1$, the measurable functions corresponding to different spectra do not differ dramatically.

Experimental measurement of the stress-relaxation behavior of human compact bone in torsion gives rise to the results which cannot be readily modeled using only previously published spectra. The slope of the relaxation curve is seen to increase in magnitude with the logarithm of time. The relaxation curve, further, is much broader than would be expected if the spectrum contained a single relaxation time. This suggests that the corresponding relaxation spectrum increases with log τ also. Therefore, the dynamic and relaxation functions corresponding to the "triangle spectrum",

$$\underset{T[\tau_1,\tau_2]}{H(\tau)} \equiv \begin{Bmatrix} \log\tau & \tau_1 < \tau < \tau_2 \\ 0, & \tau > \tau_2,\ \tau < \tau_1 \end{Bmatrix} \tag{25}$$

have been developed by Lakes and Katz[31] and utilized in the construction of a constitutive model for compact bone. The step-by-step mathematical details of this development are presented in the aforementioned article and will be omitted here.

In order to formulate a constitutive equation, the objective is to find a spectrum such that the corresponding G(t), $G'(\omega)$ and $G''(\omega)$ fit the experimental results. A first step in doing this is to obtain an approximate spectrum for the region of "small" strain using the approximations,

$$H(\tau) \cong -\left.\frac{dG(t)}{d\log t}\right|_{t=\tau},\quad H(\tau) \cong \left.\frac{2}{\pi}G''(\omega)\right|_{\omega=1/\tau} \tag{26}$$

The resulting curve is then fitted with empirical spectra for which the associated measured functions are known, and the parameters varied until a close approximation to the data can be obtained. The spectrum thus obtained is

$$\frac{H(\tau)}{G_{STD}} = 0.00318\underset{T[1,10^6]}{H(\tau)} + 0.006\underset{B[10^{-5},10^2]}{H(\tau)} + 0.002\underset{B[10^{-5},10^{-3}]}{H(\tau)} + 0.004\delta(\tau - 0.2)\tau \tag{27}$$

where the times are in seconds, and $G_{STD} = 0.590 \times 10^6$ lb/in^2 = 4.068 GN/m^2. The corresponding relaxation function is

$$\frac{G_0(t)}{G_{STD}} = 0.00318\underset{T[1,10^6]}{G(t)} + 0.006\underset{B[10^{-5},10^2]}{G(t)} + 0.002\underset{B[10^{-5},10^{-3}]}{G(t)} + 0.004e^{-t/0.2} + 0.692 \tag{28}$$

where G_T is the relaxation function associated with a triangle spectrum, Equation 25. The third term in each of these expressions contributes to the loss modulus above 100 Hz. While this region was not studied in the experiments performed by Lakes, Katz and Sternstein,[29] other data obtained using dry canine radial bone 14 indicated that the torsional loss tangent in the domain 370 Hz to 2500 Hz lies between 0.016 and 0.019. Although wet human bone would not necessarily behave in the same way, these results must suffice as a first approximation until more data are available. An increase in the loss tangent for human bone above

100 Hz is also suggested by the results of low-temperature torsional experiments upon human bone below 100 Hz described earlier. Since bone is thermorheologically complex, a stronger statement than this cannot be justified.

Now the nonlinearities observed in the present study of the torsional response of bone are primarily of a strain-dependent type. An equation of nonlinear superposition is therefore proposed to describe this response.

$$\sigma(t) = \int_{-\infty}^{t} [G(t - \tau, \epsilon(\tau)) - Ge] \frac{d\epsilon}{d\tau} d\tau + Ge \tag{29}$$

An equation similar to this has been used to describe the behavior of soft tissue by Fung[66] and is also of use in describing synthetic polymers. Since the relaxation curves obtained for human bone in the present experiments are parallel within the experimental error, the kernel may be separated:

$$G(t', \epsilon) = G_0(t') \cdot A(\epsilon) \tag{30}$$

For the present data,

$$A(\epsilon) = a_1 - a_2 e^{-(a_3 \epsilon)^2} \tag{31}$$

where $a_1 = 1.055$, $a_2 = 0.07$, $a_3 = 550$. The kernel in the constitutive Equation 29 becomes:

$$G(t,\epsilon) = [0.00318 \underset{T[1,10^6]}{G(t)} + 0.006 \underset{B[10^{-5},10^2]}{G(t)} + 0.002 \underset{B[10^{-5},10^{-3}]}{G(t)} + 0.004 e^{-t/0.2} + 0.692] [a_1 - a_2 e^{-(a_3\epsilon)}] G_{STD} \tag{32}$$

Equations 29 and 32 are supported by experimental data in the domains $1.6 \cdot 10^{-3} \leq t \leq 10^5$ sec, $3.4 \times 10^{-5} \leq \epsilon \leq 1.7 \times 10^{-3}$. While it is likely that these equations are valid for smaller strains than the above minimum, a breakdown of the formulation is to be expected at larger strains, since no provision is made for yield or fracture behavior.

Recent torsional creep experiments performed on bovine compact bone for periods of time exceeding 2×10^6 sec, indicate that very pronounced viscoelastic effects, including compliance changes exceeding a factor four, occur in bone subjected to prolonged loading.[37] When appropriate experiments of this type are performed upon human bone, Equation 32 must be modified to account for the long-term viscoelastic response.

As described earlier in Section II, some nonlinear materials can respond in a very complex fashion to details of the loading history. The simplest test for such history-dependent effects is a comparison of relaxation and recovery. Such tests were performed on bone by Lakes, Katz, and Sternstein.[29,30] Deviations from the response predicted by Equations 29 and 32 were small, less than 2%. The presence of a superimposed axial tensile stress also affected the results of the torsional experiments; however, again the perturbations were relatively small. Corrections to the constitutive Equation 29 may be found in Lakes and Katz.[31]

The preceding has been confined to a one-dimensional treatment of the torsional behavior of compact bone. In a three-dimensional analysis, the anisotropy of bone must be taken into account. The simplest constitutive equation which includes anisotropy is that for an anisotropic elastic solid:

$$\sigma_{ij} = C_{ijkl} \epsilon_{kl} \tag{33}$$

In view of the lack of agreement among viscoelastic measurements of the tension/compres-

Table 2
ELEMENTS OF THE ELASTIC MODULUS TENSOR FOR COMPACT BONE

Elastic constant: Reduced notation	Elastic constant: Full notation	Yoon and Katz[9] dried human femur	Lang[7] fresh bovine phalanx
C_{11}	C_{1111}	23.4	19.7
C_{33}	C_{3333}	32.5	33.4
C_{44}	C_{2323}	8.71	8.20
C_{12}	C_{1122}	9.06	10.1
C_{13}	C_{1133}	9.11	11.2
C_{66}	C_{1212}	7.17	3.80

Note: Units are in GN/m^2.

sion properties, we have not attempted to construct a viscoelastic constitutive equation for elements other than C_{2323}. Instead, compiled in Table 2 are the elements of the elastic modulus tensor C_{ijkl} obtained by ultrasonic methods by Lang[7] for bovine bone and by Yoon and Katz[9] for human bone. Since these data were obtained using an identical technique for all directions of loading, with specimens obtained from the same (respective) donor, scatter due to difference in technique and sample variation, is minimized.

VI. CONCLUSION: SIGNIFICANCE OF BONE VISCOELASTICITY

As discussed earlier, the presence of viscoelastic behavior in bone has a wide variety of consequences with respect to function in the body, diagnostic techniques, and the understanding of the micromechanics of bone. The consequences of viscoelastic effects may now be examined in more detail in light of presently available experimental and analytical results.

Experimental data for bone span a broad range in time and frequency. Therefore, it is appropriate to determine what portion of this range is relevant to the actual conditions of loading of bone in the body. Paul[68] obtained from gait experiments the dependence of joint forces upon time in normal walking. Lakes and Katz[20] subjected these data to Fourier analysis to determine what frequencies are present in walking, and obtained the results shown in Figure 9. Since the time resolution in Paul's experiments was quoted as 0.02 sec, the absence of appreciable amplitude in the actual spectrum above 50 Hz cannot be inferred from the graph in Figure 9. Indeed, Radin et al.[69] found that vibrations with frequencies as high as 2 kHz are transmitted through soft tissue to the leg bones in normal walking; however, they did not explicitly compare the amplitude of these vibrations with the amplitude at the fundamental frequency of walking (about 1 Hz). At the low-frequency end, the Fourier amplitude remains finite as one approaches zero frequency, which is consistent with the presence of a net "DC" component (i.e., a constant) stress upon bone. In summary, frequencies of interest in normal in vivo loading of bone, extend from near zero Hz, and possibly as high as 2 kHz. A more detailed discussion of the biomechanics of locomotion must be sought elsewhere.

Diagnostic devices based on measurements of the mechanical response of bones to external excitation have used frequencies of from a few hundred Hz in resonance impedance and impact-response experiments, up to 1 MHz in ultrasonic studies. The effect of bone viscoelasticity on techniques employing the lower frequencies appears to be minimal, compared to the effects of the soft tissue overlying the bone in question. For example, Orne[70] developed a mathematical model for the resonant behavior of the human ulna in transverse (bending)

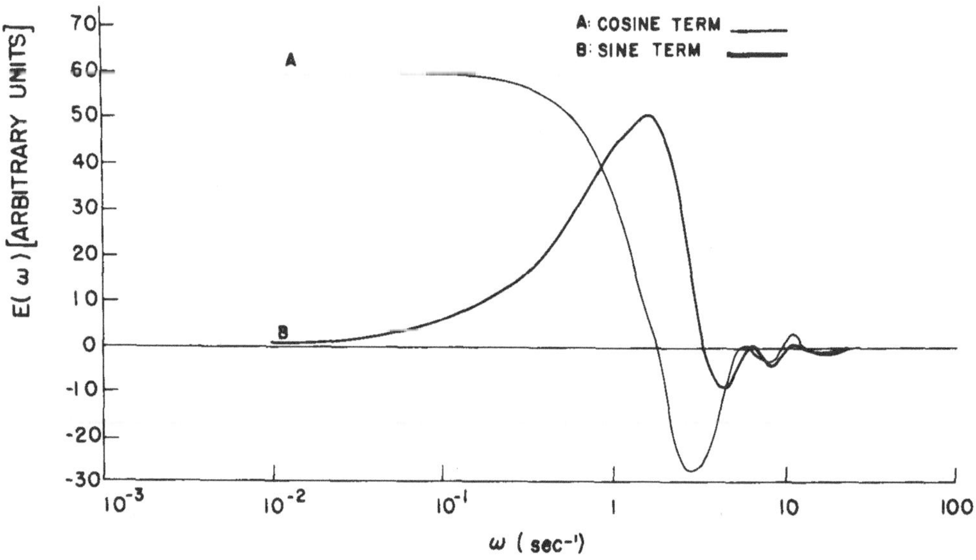

FIGURE 9. Fourier transform of the time-dependence of knee joint force in level walking, 100 paces. (From Morrison, in Lakes, R. S. and Katz, J. L., *J. Biomech.*, 7, 259, 1974.)

vibration. He found that a model based on a damping factor $\delta = 0.10$ described the resonance of an excised ulna, while a much larger damping factor $\delta = 1.0$ was necessary to describe the behavior of an ulna in vivo as a result of soft tissue viscosity. By contrast, at high frequencies the attenuation of ultrasonic waves (and hence also the loss tangent) of bone substantially exceeds that of soft tissue.[71] The attenuation in bone furthermore increases rapidly with increasing frequency, so that experimental diagnostic applications have been for the most part limited to frequencies below 0.5 to 1 MHz to ensure adequate signal strength.

Mechanical loss defined by tan δ in the linear domain is explicitly related to the coefficient of attenuation α, of elastic waves. The penetration depth ℓ is defined as the depth in a material, at which the elastic wave intensity is reduced to one-half its initial value; ℓ is given by $\ell = \log_e 2/2^{\alpha}$. Table 3 gives wave penetration depths calculated from loss tangents and attenuations obtained by various investigators.

From these data, one can conclude that the frequency at which the wave penetration depth becomes comparable to the length of a typical long bone, lies between 3.5 kHz and 500 kHz. Unfortunately, to the authors' knowledge, no direct data are available in this region. "Waves" associated with relatively low-frequency oscillations, less than 2 kHz, are not significantly attenuated in compact bone according to these data. Trabecular bone appears to offer greater damping of mechanical vibrations at frequencies near the upper end of the physiological range; this is discussed in other chapters.

What happens to the mechanical energy which is dissipated during the loading of bones in the body? Is any of this energy converted to forms which can interact with bone cells and provide a stimulus for Wolff's Law-related remodeling activity? These questions provided much of the impetus for examining relaxation mechanisms in bone. Although understanding of these mechanisms is not yet complete, the theoretical development thus far indicates that the relaxation mechanisms which have been suggested as Wolff's Law mechanisms, all can operate at physiologically relevant frequencies. The theory has therefore not ruled out any of these processes as possible Wolff's Law mechanisms by virtue of their characteristic frequency. Further testing of the various Wolff's Law hypotheses must therefore proceed along more empirical lines.

Table 3
WAVE PENETRATION DEPTHS IN CALCIFIED TISSUES

Frequency (Hz)	tan δ	α (cm^{-1})	ℓ (cm)	Bone type	Ref.
160	0.018[a]	3×10^{-5}	11,000	Wet dog radius in bending vibration	14
500	0.025[a]	1.4×10^{-4}	2,500	Wet human tibia, dynamic test, along bone axis	13
3k	0.016[a]	5×10^{-4}	690	Wet dog radius	14
3.5k	0.025[a]	9.7×10^{-4}	360	Wet human tibia	13
0.5M	0.58	2.24[a]	0.15	Compact bone	71
0.6M		1.5[a]	.077	Skull	72
0.8M	—	9[a]	.038	Skull	72
1.2M	—	17[a]	.020	Skull	72
1.6M	—	32[a]	.011	Skull	72
1.8M	—	42[a]	.0083	Skull	72
2M	0.82	12.6[a]	.028	Compact bone	71
2.25M	—	53[a]	.0065	Skull	72

[a] This indicates the quantity originally measured, from which the others have been calculated.

REFERENCES

1. **Ferry, J. D.,** *Viscoelastic Properties of Polymers*, John Wiley & Sons, New York, 1970.
2. **Gross, B.,** *Mathematical Structure of the Theories of Viscoelasticity*, Hermann, Paris, 1953.
3. **Norris, D. M.,** Propagation of a stress pulse in a viscoelastic solid, *Exp. Mech.*, 11, 297, 1967.
4. **Evans, F. G.,** *Mechanical Properties of Bone*, Charles C Thomas, Springfield, Ill., 1973.
5. **Katz, J. L. and Mow, V. C.,** Mechanical and structural criteria for orthopaedic implants, *Biomater. Med. Devices Artif. Organs*, 1 (4), 575, 1973.
6. **Reilly, D. T. and Burstein, A. H.,** The mechanical properties of cortical bone. *J. Bone Intern. Surg.*, 56A, 1001, 1974.
7. **Lang, S. B.,** Elastic coefficients of animal bone, *Science*, 165, 287, 1969.
8. **Yoon, H. S. and Katz, J. L.,** Ultrasonic wave propagation in bone. I. Theoretical considerations for hexagonal symmetry, *J. Biomech.*, 9, 407, 1976a.
9. **Yoon, H. S. and Katz, J. L.,** Ultrasonic wave propagation in human cortical bone. II, Measurements of elastic properties and microhardness, *J. Biomech.*, 9, 459, 1976b.
10. **Rauber, A.,** Elasticitat und festigkeit der knochen, *Anatomisch Physiologische Studie*, 75 Leipzig, 1876.
11. **Hancox, N.,** *Biology of Bone*, Cambridge University Press, 1972.
12. **Sedlin, E.,** A rheological model for cortical bone, *Acta Orthop. Scand.*, 36, suppl. 83.
13. **Smith, R. and Keiper, D.,** Dynamic measurement of viscoelastic properties of bone, *Am. J. Med. Elec.*, 4, 156, 1965.
14. **Thompson, G.,** Experimental studies of lateral and torsional vibration of intact dog radii, Ph.D. thesis, Biomedical Engineering, Stanford University, 1971.
15. **Bargren, J. H., Bassett, A. L., and Gjelsvik, A.,** Mechanical properties of hydrated cortical bone, *J. Biomech.*, 7, 239, 1974.
16. **Laird, G. W. and Kingsbury, H. B.,** Complex viscoelastic moduli of bovine bone, *J. Biomech.*, 6, 59, 1973.
17. **Black, J. and Korostoff, E.,** Dynamic mechanical properties of viable human cortical bone, *J. Biomech.*, 6, 435, 1973.
18. **Currey, J. D.,** The effects of strain rate, reconstruction, and mineral content on some mechanical properties of bovine bone, *J. Biomech.*, 8, 81, 1975.
19. **McElhaney, J.,** Dynamic response of bone and muscle tissue, *J. Appl. Physiol.*, 21, 1231, 1966.
20. **Lakes, R. S. and Katz, J. L.,** Interrelationships among the viscoelastic functions for anisotropic solids: application to calcified tissues and related systems, *J. Biomech.*, 7, 259, 1974.
21. **Bonfield, W. and Datta, P. K.,** Young's modulus of compact bone, *J. Biomech.*, 7, 147, 1974.

22. **Tennyson, R. C., Ewert, R., and Niranjan, V.**, Dynamic viscoelastic response of bone, *Experim. Mech.*, 12, 502, 1972.
23. **Lugassy, A. A. and Korostoff, E.**, Viscoelastic behavior of bovine femoral cortical bone and sperm whale dentin, in *Research in Dental and Medical Materials*, Plenum, New York, 1969.
24. **Lugassy, A. A.**, Mechanical and viscoelastic properties of bone and dentin in compression, Ph.D. dissertation, Metallurgy and Materials Science, University of Pennsylvania, 1968.
25. **Dempster, W. T. and Liddicoat, R. T.**, Compact bone as a non-isotropic material, *Am. J. Anat.*, 91, 3, 331, 1952.
26. **Currey, J. D.**, Anelasticity in bone and echinoderm skeletons, *J. Exp. Biol.*, 43, 279, 1965.
27. **Bonfield, W. and Li, C. H.**, Anisotropy of nonelastic flow in bone, *J. Appl. Phys.*, 38, 6, 2450, 1967.
28. **Adler, L. and Cook, K. V.**, Ultrasonic parameters of freshly frozen dog tibia, *J. Acoust. Soc. Am.*, 58, 1107, 1975.
29. **Lakes, R. S., Katz, J. L., and Sternstein, S.**, Torsional dynamic and relaxation properties of human and bovine cortical bone, *Proc. 21st Ann. Orthop. Res. Soc. Meet.*, San Francisco, 1975.
30. **Lakes, R. S., Katz, J. L., and Sternstein, S.**, Viscoelastic properties and behavior of cortical bone: Part I: torsional and biaxial studies, *J. Biomech.*, 12, 657, 1979.
31. **Lakes, R. S. and Katz, J. L.**, Viscoelastic properties and behavior of cortical bone: part III: a non-linear constitutive equation, *J. Biomech.*, 12, 689, 1979.
32. **Smith, J. W. and Walmsley, R.**, Factors affecting the elasticity of bone, *J. Anat.*, 93, 503, 1959.
33. **Frost, H.**, Measurement of osteocytes per unit volume and volume components of osteocytes and canaliculi in man, *Henry Ford Hosp. Med. Bull.*, 8, 208, 1960.
34. **Evans, F. G. and Bang, S.**, Physical and histological differences between human fibular and femoral compact bone, in *Studies on the Anatomy and Function of Bones and Joints*, Evans, F. G., Ed., Springer Verlag, Berlin, 1966, 142.
35. **Black, J., Ph.D.**, Dissertation, Department of Metallurgy and Material Science, University of Pennsylvania, Philadelphia, 1972.
36. **Fitzgerald, E. R.**, Dynamic mechanical measurements during the life to death transition in animal tissues, *Biorheology*, 12, 397, 1975.
37. **Lakes, R. S. and Saha, S.**, Torsional deformation of bone subjected to prolonged loading, *Proc. 24th Ann. Orthop. Res. Soc. Meet.*, Dallas, 1978.
38. **Green, E. and Rivlin, R. S.**, The mechanics of non-linear materials with memory, *Arch. Rational Mech. Anal.*, 1, 1, 1957.
39. **Ward, I. M. and Onat, E. T.**, Non-linear mechanical behavior of oriented polypropylene, *J. Mech. Phys. Solids*, 11, 217, 1963.
40. **Pipkin, A. C. and Rogers, T. G.**, A non-linear integral representation for viscoelastic behavior, *J. Mech. Phys. Solids*, 18, 59, 1968.
41. **Lakes, R. S. and Katz, J. L.**, Viscoelastic properties and behavior of cortical bone. part II. Relaxation mechanisms, *J. Biomech.*, 12, 679, 1979.
42. **Lockett, F. J. and Gurtin, M. E.**, Frequency response of non-linear viscoelastic solids, *Brown University technical report, NONR 562 (10), NONR 562(30)*, 1964.
43. **Tischendorf, F.**, Das verhalten der haversschen systeme bei belastung, *Roux' Archiv Fur Entwicklungs-mechanik*, 145, 318, 1951.
44. **Mouradian, W. E.**, Electrical response of wet bone, M.S. thesis, Massachusetts Institute of Technology, Department of Metallurgy and Materials Science, 1973.
45. **Renton, D.**, The viscoelastic properties of bone, M.A. Sc. thesis, University of Waterloo, Ontario, 1970.
46. **Bassett, A.**, Electrical effects in bone, *Sci. Am.*, 213 (4), 18, 1965.
47. **Nye, P.**, *Physical Properties of Crystals*, Oxford, Clarendon, 1957.
48. **Liboff, A. and Shamos, M. H.**, Solid state physics of bone, in *Biological Mineralization*, Zipkin, I., Ed., John Wiley & Sons, New York, 1973.
49. **Hamill, D. and Harper, R. A.**, Private communication, 1975.
50. **Egan, E., Jr., Wakefield, Z., and Elmore, K.**, Low temperature heat capacity and entropy of hydroxyapatite, *J. Am. Chem. Soc.*, 73, 5579, 1951.
51. **Liboff, A., Shamos, M. H., and deVirgilio, W.**, The piezoelectric modulus of bone, Paper Th AM-f3, 15th Ann. Meet. Biophys. Soc., New Orleans, 1971.
52. **Marino, A., Becker, R., and Backman, C.**, Dielectric determination of bound water in bone, *Phys. Med. Biol.*, 12, 3, 367, 1967.
53. **Zener, C.**, Internal friction in solids — general theory of thermoelastic internal friction, *Phys. Rev.*, 53, 90, 1938.
54. **Ascenzi, A. and Bonucci, E.**, The compressive properties of single osteons, *Anat. Rec.*, 161, 337, 1966.
55. **Bur, A. J.**, Measurements of the dynamic piezoelectric properties of bone as a function of temperature and humidity, *J. Biomech.*, 9, 495, 1976.

56. **Anderson, J. C. and Eriksson, D.,** Piezoelectric properties of dry and wet bone, *Nature*, 227, 491, 1970.
57. **Nowinski, J. L. and Davis, C. F.,** The flexure and torsion of bones viewed as anisotropic poroelastic bodies, *Int. J. Eng. Sci.*, 10, 1063, 1972.
58. **Rusch, K.,** Dynamic behavior of flexible open cell foams, Ph.D. thesis, University of Ohio, Department of Polymer Science, 1965.
59. **Gent, A. N. and Rusch, K. C.,** Viscoelastic behavior of open cell foams, *Rubber Chem. Tech.*, 39, 389, 1966.
60. **Piekarski, K.,** Fracture of bone, *J. Appl. Phys.*, 41 (1), 215, 1970.
61. **Zener, C.,** Theory of the elasticity of polycrystals with viscous grain boundaries, *Phys. Rev.*, 60, 906, 1941.
62. **Ke, T.,** Experimental evidence of the viscous behavior of grain boundaries in metals. *Phys. Rev.*, 71, 533, 1947.
63. **Lakes, R. S. and Saha, S.,** Cement Line Motion in Bone, *Science*, 204, 501, 1979.
64. **Lakes, R. S., Katz, J. L., and Harper, R. A.,** Dielectric relaxationin cortical bone, *J. Appl. Phys.*, 48, 808, 1977.
65. **Tobolsky, A. B.,** *Properties and Structure of Polymers*, John Wiley & Sons, New York, 1960.
66. **Fung, Y. C.,** Stress-strain-history relations of soft tissues in simple elongation, in *Biomechanics, Its Foundations and Objectives*, Fung, Y. C., Perrone, N., and Anliker, M., Eds., Prentice Hall, New York, 1972.
67. **Lowan, A. N.,** *Tables of Sine, Cosine and Exponential Integrals*, National Bureau of Standards Computation Lab, 1940.
68. **Paul, J. P.,** Load actions on the human femur in walking and same resultant stresses. *Exp. Mech.*, 11, 121, 1971.
69. **Radin, E. L., Parker, M. G., Pugh, J. W., Steinberg, R. S., Paul, I. L., and Rose, R. M.,** Response of joints to impact loading. III. Relationship between trabecular microfractures and cartilage degeneration, *J. Biomech.*, 6, 51, 1973.
70. **Orne, D.,** The *in vivo* driving point impedance of the human ulna — a viscoelastic beam model, *J. Biomech.*, 7, 249, 1974.
71. **Brown, S. A. and Mayor, M. B.,** Ultrasonic assessment of early callus formation, *Biomed. Eng.*, 11, 124, 1976.
72. **Goldman, D. E. and Hueter, T. F.,** Tabular data of the velocity and absorption of high frequency sound in mammalian tissues, *J. Acoust. Soc. Am.*, 28, 35, 1956.
73. **Lakes, R. S.,** Ph.D. dissertation, Department of Physics, Rensselaer Polytechnic Institute, Troy, New York, 1975.

Chapter 6

MECHANICAL PROPERTIES OF CANCELLOUS BONE

R. Van Audekercke and M. Martens

TABLE OF CONTENTS

I. INTRODUCTION

Cancellous (trabecular, spongy) bone is structurally an open cell foam which is present at the epiphyseal and metaphyseal region of long bones and within the cortical confinements of flat and short bones (Figure 1). Trabecular or cancellous bone is continuous with the inner surface of the cortical shell and presents a three-dimensional lattice composed of plates and columns of bone.

The mechanical properties of cancellous bone have been studied less thoroughly than those of cortical bone. Topography among and within bones is an important variable to be considered for the determination of density, trabecular contiguity, and mechanical properties of cancellous bone (Figure 2). A comprehensive review of previous studies of mechanical properties of trabecular bone yields the following data.

II. ELASTIC PROPERTIES

A. Modulus of Elasticity

The modulus of elasticity has been determined by several authors. Table 1 lists the values for this parameter given in the literature. The high values McElhaney and Byars[10] obtained for cancellous bone of vertebral bodies are in contrast with other authors. The large standard deviation McElhaney found in the values for this elastic constant can be attributed to the wide age range (1 month to 84 years of age). Evans and King[5] recorded for cancellous bone from various parts of the femur, testing rectangular and cubic specimens of embalmed bone in compression; the highest mean modulus of elasticity was in the femoral neck, followed in descending order by specimens from the head, lateral condyle, medial condyle, and greater trochanter (one specimen only). Kniese[8] determined the modulus of elasticity of tibial condylar cubes, loaded in different directions. He noted higher values for the longitudinally loaded specimens than for the transversely loaded specimens (Table 1). The condition of the test specimens is not actually stated, and only one specimen for each direction has been tested.

Carter and Hayes[3] investigated the relationship between apparent density of trabecular bone specimens and compressive modulus. The authors concluded that the modulus was proportional to the cube of the apparent density. They also examined the influence of bone marrow in the pores of cancellous bone on the mechanical behavior of the test specimens and also the influence of strain rates (Table 2). Bone marrow did not influence the stiffness of the trabecular bone specimens except for the higher strain rates (10 per sec).

In these experiments the influence of bone marrow was significant for the higher strain rates. However, the plate pores of the testing apparatus were much smaller than those of trabecular bone and the influence of marrow on the recorded mechanical properties can be due to restricted marrow flow through the plates of the testing instrument rather than through the pores of the trabecular bone specimens.

Townsend, Rose, and Radin[14] reported the modulus of elasticity of single trabeculae by buckling studies of single trabeculae from the subchondral region of the human medial tibial plateau. By extrapolation to ideal slenderness ratios the authors obtained a value for Young's Modulus reasonably similar to cortical bone.

The comments given in Table 1 demonstrate that condition of test specimens and strain rate were not uniform for the various studies. This and the fact that some authors only tested a limited number of specimens from different areas within a bone specimen does not allow firm conclusions with regard to this mechanical parameter. The same restrictions hold for compressive strength of cancellous bone.

B. Poisson's Ratio

McElhaney[10] determined Poisson's ratio from specimens of twenty-eight vertebral bodies

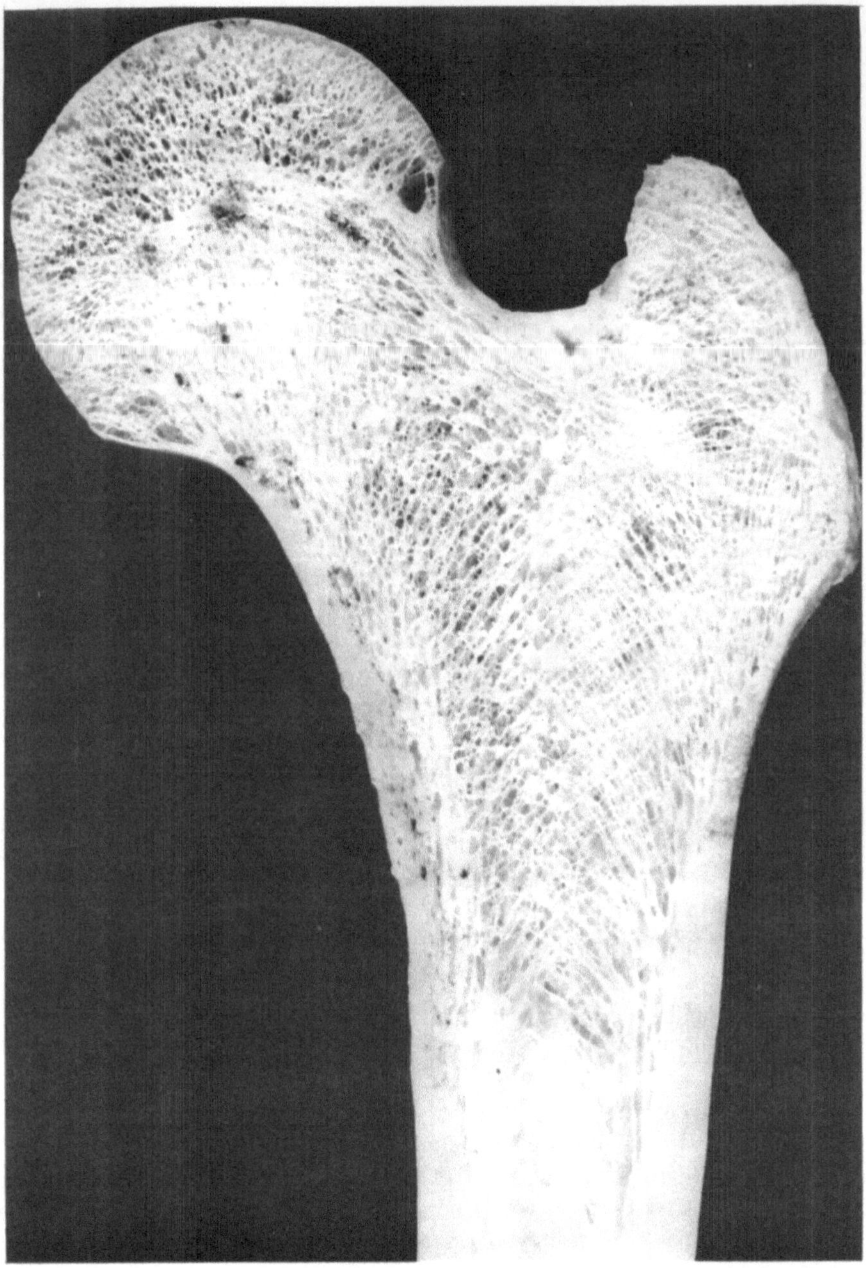

FIGURE 1. A longitudinal section at the proximal femur reveals the absence of cancellous bone at the shaft (bottom) and the tapering of the cortical wall at the metaphyseal region. At the femoral head the cortical bone represents only a thin shell continuous with the underlying trabecular network

from seven individuals between 45 and 79 years of age. They obtained a value for this elastic constant of 0.14 ± 0.09.

III. ULTIMATE PROPERTIES

A. Tensile Strength

Sonoda[13] determined the tensile properties of cancellous bone from fresh moist specimens

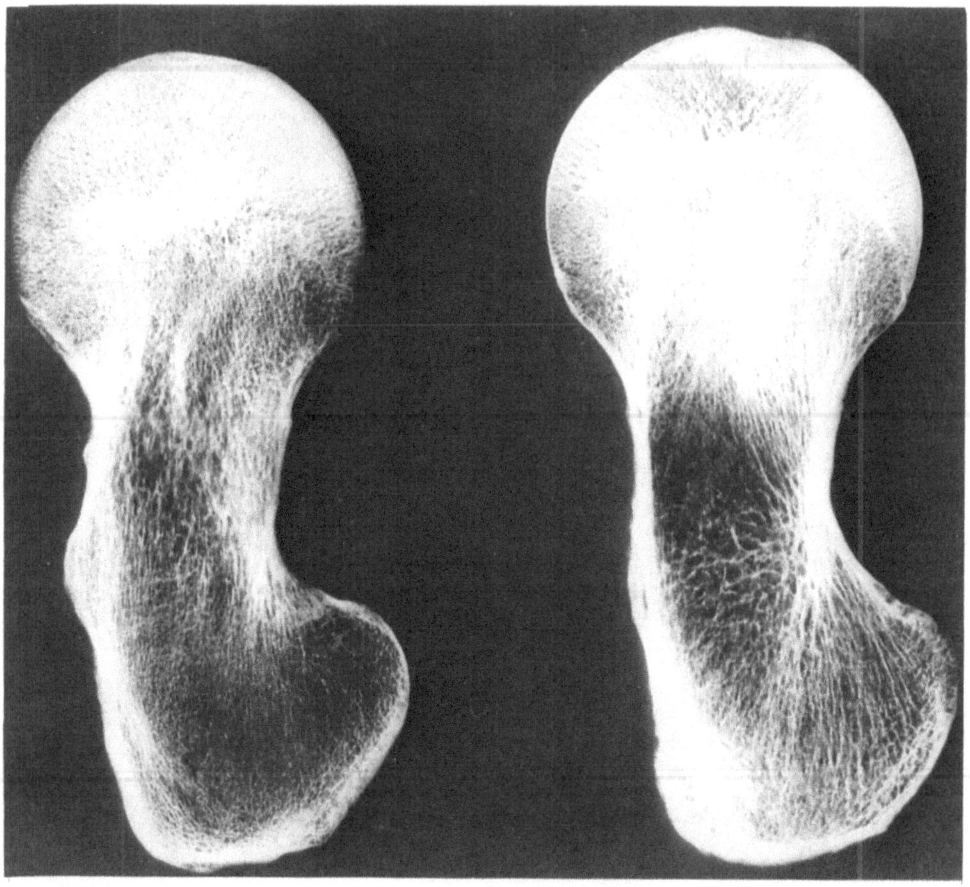

FIGURE 2. A radiograph of bone slabs taken through the femoral head, neck, and intertrochanteric region demonstrates the variation in density and contiguity of the trabecular bone tissue at these different regions.

from thoracic and lumbar vertebrae of people 30 to 39 years of age. The value obtained for tensile strength: $1.2/0.1 \times 10^6$ N/m^2.

B. Compressive Strength

The values of compressive strength mentioned by different authors are given in Table 2. Because density and structural orientation of cancellous bone varies considerably between different bones and even within each bone, it is mandatory to refer for the mechanical properties of cancellous bone to the bone tested indicating the precise site and orientation of the test specimen.

1. Vertebral Bodies

McElhaney[10] found a lower mean of compressive strength for cancellous bone of vertebral bodies than Sonoda,[13] although it is still higher than that reported by Yokoo[16] and Galante et al.[6] The orientation of the axis of the specimens was always parallel to the height of the vertebral body. The relation of compressive strength of human cancellous bone from vertebral bodies to its density has been investigated by Galante et al.[6]

Two types of density are considered in these experiments. Real density is computed by dividing wet weight by volume of bone matrix as determined by water displacement. Real density of cancellous bone is a function of its composition. An increase in mineral content results in an increase in real density. A significant difference in compressive strength as a

Table 1
E MODULUS OF CANCELLOUS BONE

Specimen (no. of specimens)	E Modulus (10^6 N/m^2)		Comments	Ref.
Lumbar vertebral bodies			Wet specimens.	16
40—49 years	88			
60—69 years	69			
Tibial condyle			Condition of specimen not actually stated.	8
longitudinal direction	61		One specimen from each direction obtained	
transverse direction	51		from one femur and one tibia.	
Femoral head				
anteroposterior	57			
direction of the neck	87			
Femoral head (19)	581	± 17	Embalmed adult human femur.	5
Lateral femoral condyle (13)	571	± 47	Wet rectangular prisms and cubic specimens	
Medial femoral condyle (23)	418	± 31	Long axis of specimens cut in various directions.	
Greater trochanter (1)	81			
Femoral neck (5)	780	± 41	We assume that the values published by Evans [5] and expressed in kg/mm^2 mean in fact kg/cm^2. Otherwise these values would be two orders of magnitude too high.	
Vertebral bodies (288)	1518	± 1172	Wet specimens one month to 89 years of age.	10
	Mean	**S.D.**		
Femoral head (30)	344	27,6	Osteoarthrotic femoral heads. Slow strain rates (0,004). 30 specimens from 5 femoral heads. Cylindrical specimens and long axis parallel to the long axis of the femoral neck.	12
Vertebral bodies				
males	55,6	0,7	Dry defatted specimen	9
females	35,1	0,6	Slow deformation rate	
Proximal bodies				
males	34,6	0,4		
females	23,1	0,4		
Tibial plateau (100)				
0,001[a]	56,6	9,7	Wet specimens.	3
0,01	75,5	11,8	Varying strain rates.	
0,1	81,5	8,0	Specimen with and without marrow in situ.	
1	81,2	17,1		
10	83,7[b]	13,8[b]		

[a] Strain rate (per second).
[b] Figure for E at strain rate 10/sec only for specimens without marrow in situ.

function of the real density was not present in his results. Apparent density is calculated by dividing the wet weight by the total specimen volume. The percent porosity of the specimens is given by dividing the apparent density by the real density and multiplying the quotient by 100. Ash weight per specimen was likewise computed. Galante[6] found significant relations between compressive strength and apparent density of their specimens. Compressive strength varied directly with apparent density, and a straight line could be fitted by the method of

Table 2
COMPRESSIVE STRENGTH OF CANCELLOUS BONE

Specimen	Compressive str. ($\times 10^6$ N/m²)	Comments	Ref.
Vertebral bodies			
40—49 years	1.86	Number of specimens not mentioned.	16
60—69 years	1.37	Slow strain rate.	
Femoral head			
anteroposterior direction	11.4	Condition of specimens not stated.	8
direction of neck	9.4		
Tibial plateau			
longitudinal	5.8	Only one specimen from one bone	
transverse	2.8	for each direction.	
Femoral head	7.61 ± 2.25	Embalmed bone.	5
Lateral tibial condyle	5.05 ± 4.30	Slow strain rate.	
Medial tibial condyle	4.65 ± 2.95		
Greater trochanter	1.35 ± 0.66		
Femoral neck	8.75 ± 4.57		
Cervical spine (20—39 yr)	12.5 ± 0.2	Slow strain rate.	13
Upper thoracic (20—39 yr)	8.63	Number of specimens not mentioned.	
Middle thoracic (20—39 yr)	7.7	Young age group.	
Lower thoracic (20—39 yr)	7.2 ± 0.1		
Lumbar (20—39 yr)	6.3 ± 0.1		
Vertebral bodies under 50 yr	4.25	Fresh specimens.	15
Vertebral bodies over 50 yr	2.54	Low strain rate.	
Calcaneus under 50 yr	3.91		
Calcaneus over 50 yr	3.55		
Vertebral bodies	4.1 ± 3.4	288 specimens and age variation between 1 month and 89 yr. Slow strain rate.	10
L3 or L4 vertebral bodies	2.06 ± 0.25	72 wet specimens. Slow strain rate.	6
Femoral head	0.25 to 13.5	Osteoarthrotic femoral heads removed during arthroplasty. Slow strain rate.	12
Femoral condyles	15.9 to 25.8	Mean values for different locations at the femoral condyle.	2
Tibial plateaus	13.9 to 23.2	Mean values for different locations at the tibial plateaus	

Specimen	Mean	S.D.	Comments	Ref.
Vertebral bodies				
males	4.6	0.3	Dry defatted specimens from	9
females	2.7	0.2	64 autopsy subjects.	
Proximal tibia				
males	3.9	0.3	Slow deformation rate	
females	2.2	0.2	(0.05 mm/min).	

Table 2 (continued)
COMPRESSIVE STRENGTH OF CANCELLOUS BONE

Specimen	Compressive str. (× 10^6 N/m²)		Comments	Ref.
Proximal tibia				3
0.001[a]	4.2	0.6	Wet specimens with and without marrow.	
0.01	4.1	0.7		
0.1	5.8	0.7	One hundred cylindrical specimens.	
1	6.7	0.8	Varying strain rate.	
10	9.13	1.32[b]		

[a] Strain rate (per second).
[b] Figure for strain rate 10/sec only for specimens without marrow.

least squares with a 0.01 significance. The same correlation was found between dry weight and ash weight as related to total specimen volume. However, apparent density is a more realistic entity and revealed less variation around predicted regression lines. The apparent density changes are mainly due to changes in volume of the void spaces. Minor variations in porosity induce major variations in apparent density. Composition of the matrix is the other factor which determines apparent density, but because of the high porosity variations in matrix density, variations in matrix composition are shaded by variations in pore volume and serve only a minor contribution upon the apparent density.

Weaver and Chalmers[15] examined the compressive strength of vertebral bodies and the calcaneus by testing cubes of fresh unembalmed cancellous bone at slow strain rates. The average calcaneal strength and ash weight were significantly higher than the corresponding vertebral values (Table 2). Their results also show that the compressive strength of cancellous bone is closely related to mineral content in both the vertebra and the calcaneus, as indicated by the correlation coefficients ranging from 0.701 to 0.843 for the different subgroups (female, male, over and under fifty years of age).

Lindahl[9] reported on compressive strength of cancellous bone from vertebral bodies and proximal tibia (Table 2). The author also investigated the relationship between apparent density and compressive strength of spongy bone. These results confirmed the close correlation between compressive strength and apparent density (correlation coefficient of 0.7). His data are based upon testing dry defatted bone specimens at slow strain rates.

2. Cancellous Bone From Long Bones

Data on the mechanical properties of cancellous bone from long bones are not as abundant as those for vertebrae. Hardinge[7] determined the force required to punch out little round pieces of cancellous bone from cross sections at the femoral neck and head of ninety-four femurs from embalmed cadavers. Values were reported on the average force required to crush the specimens. Strength as force per unit area was not mentioned. This study revealed large topographic variations for crushing force at the femoral head and neck (ratio 1 to 3 for different areas). Furthermore, his data demonstrate a bandlike region of higher crushing force extending from the superior aspect of the head to the inferior aspect of the neck along the dense vertical trabecular system in the upper femur. Knese[8] reported regional and directional differences in compressive strength of cancellous bone of a femur. However the interpretation of his results is difficult, since only one specimen from each region was tested while the loading direction was changed for the different specimens. He also did not state the condition of the bone specimens. These results therefore are not listed in the table. Differences in the compressive properties of standardized specimens of cancellous bone from various parts of the femur have been studied by Evans and King.[5] They tested rectangular prisms with a slenderness ratio of 11/4 and small cubes (0.79 cm) from embalmed cadavers.

The authors found the greatest mean compressive strength for specimens from the femoral neck, followed in descending order by specimens from the femoral head, the lateral condyle, medial condyle, and greater trochanter region. Specimens from the femoral head also absorbed the most energy to failure, followed in descending order by specimens from the femoral neck, lateral condyle, medial condyle, and greater trochanter. The number of specimens for some regions was rather small, and the authors used embalmed bone and slow strain rates for their experiments. Therefore firm conclusions for regional variations in mechanical properties for femoral cancellous bone cannot be drawn from these experiments. Ducheyne et al.[4] reported, contrary to the previous authors, higher compression strength of cancellous bone at the medial femoral condyle than in the lateral condyle. This reversed order correlates better with the functional loading at the knee joint where the larger part of the joint reaction force during walking is borne by the medial compartment of the knee.

Schoenfeld et al.[12] also investigated the compressive strength of human cancellous bone derived from femoral heads. They found a wide range from 1.48 to 13.8 kg/cm^2 for this value. A correlation coefficient of 0.7 related the compressive strength to the apparent density of cancellous bone. The long axis of the test specimens was oriented parallel to the long axis of the femoral neck, and a slow strain rate on an Instron® Testing Instrument (2.5 mm/min) was applied. They used osteoarthrotic bone specimens from five surgically removed femoral heads during an arthroplasty. The use of pathological bony tissue could explain the large variation for the compressive strength they obtained for the different specimens. In contrast with Hardinge's work they also concluded that the relatively weak and strong regions of cancellous bone at the femoral head appear to occur on a random basis. Their results are not relevant for mechanical behavior of healthy cancellous bone at the femoral head because of the use of osteoarthrotic femoral heads.

Behrens et al.[2] studied the variations in strength and structure of cancellous bone at the femoral condyles and tibial plateau. The sites chosen for the bone specimen were at the joint contact areas at 0°, 45°, and 90° of flexion. Fresh bone was obtained from 10 autopsy subjects ranging in age from 40 to 92 years.

Compressive strength of cancellous bone 5 mm below the cortical surface was determined by loading bone specimens using a circular indentor in an Instron® Testing Machine at low strain rate (10 cm/min). Tibial cancellous bone displayed about the same strength as femoral bone. The average figures showed a variation in bone strength with the location in relation to the joint. On the lateral femoral condyle cancellous bone at joint contact area at 0° and 45° flexion was strongest while on the medial condyle bone specimens at 45° flexion yielded the highest values.

For both the femur and the tibia, the cancellous bone on the medial side showed higher values for compressive strength than on the lateral side. Bulk specimen density (dry bone weight/total specimen volume) and linear absorption coefficient of bone per cm (μB) via radiological absorption were found to correlate with strength but did not fully account for the wide strength variations. The authors stated that trabecular organization reflected by the morphology of the trabecular pattern and fractional areas (area of bone/area of specimen across a section parallel to the surface) could be the most important single factor. Also other authors as Pugh et al.[11] demonstrated the relationship between mechanical properties and microstructure of cancellous bone at the femoral condyle. These authors concluded that variation in stiffness of subchondral bone at the femoral condyle for different specimens is mainly the result of the variation in the contiguity factor (C_t). The trabecular contiguity (C_t) is a stereological concept that enables the quantitation of the spatial geometry or degree of uniformity of a structural network. The higher the C_t the more perfect the network. C_t has limits between 0.5 and 1. C_t equal to 0.5 corresponds to a system of parallel sheets with no interconnections, and C_t equal to 1 corresponds to a perfect network of trabeculae. The contiguity factor and the stiffness of bone showed statistically significant differences. Other microstructural concepts such as Volume Fraction Bone (volume bone/total volume), average trabecular thickness, and trabecular density (number of trabeculae per unit length) did not exhibit statistically significant differences. The relationship between stiffness and contiguity factors of cancellous bone is based upon the fact that trabecular bone can be seen as a structure of interconnected beams and plates, and the most effective way to increase stiffness of a beam is not by making it more massive but by supporting it near its center with a column, that is to connect it with the rest of the structure at one more point.

3. Viscoelastic Behavior

Carter and Hayes[3] conducted compression tests on 100 trabecular bone specimens removed from tibial plateaux (5 mm thick and 10.3 mm radius). They investigated the influence of strain rate (Table 2) and presence of bone marrow. Specimens with and without bone marrow had similar compressive strength except at a strain rate of 10/sec. This could be explained by their testing conditions (cfr. comment for E modulus of these authors). Strength seemed to be proportional to strain rate raised to the 0.06 power. The authors noted that strength was proportional approximately to the square of the apparent density.

Galante et al.[6] also demonstrated the time dependency of compressive strength of cancellous bone at deformation rates of 0.01 and 1 cm/min.

4. Aging

The aging effect on the compressive strength of cancellous bone has been studied by Weaver and Chalmers[15] (Table 2). They tested cancellous bone specimens from vertebral bodies of 137 cadavers and from the calcaneus of 99 subjects. Their tests revealed an obvious trend towards decreasing bone strength in compression and mineral content with advancing age, and this holds especially for vertebral bodies. The diminution appears earlier in females and is more profound. As a result of this phenomenon, strength and mineral content are significantly lower in female cadavers in comparison with the male group, although for the age group below fifty there is no significant difference between both sexes.

Evans and King[5] determined the compressive properties of cancellous bone from adult human femora of ten individuals varying from 45 to 88 years of age. A clear tendency toward a decrease in maximum compressive stress and energy absorbed to failure was observed after 70 years of age. Age changes in the compressive properties of cancellous vertebral bone were also investigated by Bartley et al.[1] using forty lumbar bodies. They noticed a striking similarity between the distribution of the ash content vs. age and the compressive strength vs. age.

IV. CONCLUSIONS

Cancellous bone is an open-cell porous structure. The varying porosity (range from 30% to more than 90%) among and within bones makes topography of the test specimens an important variable in the determination of mechanical properties of cancellous bone.

The porosity of normal bone tissue shows an approximately linear relationship to apparent density and ash density. Therefore porosity, ash density, and apparent density are reasonable measures of the amount of mineralized tissue. A relationship between apparent density and mechanical properties (E modulus, compressive strength) has been shown by several authors. The orientation and pattern of the trabecular network also changes between and within the bone structures. This results in anisotropy, and description of the microstructure of cancellous bone at various regions becomes important. A stereological concept, trabecular contiguity (C_t), enables the quantitation of spatial geometry. There seems to be a relationship between stiffness and contiguity factor of cancellous bone. An influence of strain rate on mechanical behavior has been noticed, but this is relatively small compared with the influence of apparent density. Presence of bone marrow does not affect mechanical behavior at strain rates up to 1/sec. Aging effects have been found for trabecular bone of vertebral bodies, femur, and calcaneus. The decrease in maximum compressive strength seems to appear earlier and to be more profound in females.

REFERENCES

1. **Bartley, M. H., Arnold, J. S., Haslam, R. K., and Jee, W. S. S.,** The relationship of bone strength and bone quality in health, disease and aging. *J. Gerontol.*, 21, 517, 1966.
2. **Behrens, J. C., Walker, P. S., and Shoje, H.,** Variations in strength and structure of cancellous bone at the knee. *J. Biomech.*, 7, 201, 1974.
3. **Carter, D. R. and Hayes, W. C.,** The compressive behavior of bone as a two phase porous structure. *J. Bone J. Surg.*, 59-A, 954, 1977.
4. **Ducheyne, P., Heymans, L., Martens, M., Aernoudt, E., De Meester, P., and Mulier, J. C.,** The mechanical behaviour of intracondylar cancellous bone of the femur at different loading rates. *J. Biomech.*, 10, 747, 1977.
5. **Evans, F. G. and King, A. I.,** Regional differences in some physical properties of human spongy bone, in *Biomechanical Studies of the Musculoskeletal System*, Evans, F. G., Ed., Charles C Thomas, Springfield, 1961, 49.
6. **Galante, J., Rostoker, W., and Ray, R. D.,** Physical properties of trabecular bone. *Calcif. Tissue Res.*, 5, 236, 1970.
7. **Hardinge, M. G.,** Determination of the strength of the cancellous bone in the head and neck of the femur. *Surg. Gynecol. Obstet.*, 89(4), 439, 1949.
8. **Knese, K. H.,** Knochenstruktur als Verbundbau.Versuch einer technischen Deutung der Materialstruktur des Knochens, in *Zwangslose Abhandlungen aus dem Gebiet der Normalen und Pathologischen Anatomie*, Bargmann, W. and Doerr, W., Eds., Georg Thieme, Stuttgart, 1958, no. 4.
9. **Lindahl, O.,** Mechanical properties of dried defatted spongy bone. *Acta Orthop. Scand.*, 47, 11, 1976.
10. **McElhaney, J. H. and Byars, E. F.,** Dynamic response of biological materials. American Society of Mechanical Engineers Publ. 65-WA/HUF 9, 1, 1965.
11. **Pugh, J. W., Rose, R. M., and Radin, E. L.,** A structural model for the mechanical behaviour of trabecular bone. *J. Biomech.*, 6, 657, 1973.
12. **Schoenfeld, C. M., Lautenschlager, E. P., and Meyer, P. R.,** Mechanical properties of human cancellous bone in the femoral head. *Med. Biol. Eng.*, 12, 313, 1974.
13. **Sonoda, T.,** Studies on the strength for compression, tension and torsion of the human vertebral column. *J. Kyoto Pref. Med. Univ.*, 71(9), 659, 1962.
14. **Townsend, P. R., Rose, R. M., and Radin, E. L.,** Buckling studies of single human trabeculae. *J. Biomech.*, 8, 199, 1975.
15. **Weaver, J. K. and Chalmers, J.,** Cancellous bone: its strength and changes with aging and an evaluation of some methods for measuring its mineral content. *J. Bone J. Surg.*, 48-A, 289, 1966.
16. **Yokoo, S.,** Compression test of the cancellous bone. *J. Kyoto Pref. Med. Univ.*, 51(3), 273, 1952.

Chapter 7

FRACTOGRAPHY OF BONE

K. Piekarski

TABLE OF CONTENTS

I. INTRODUCTION

Fractographic aspects of bone have always been examined in the same manner as for other engineering materials. The great difference is the fact that bone is a living tissue.

The activity of the cell controls its size, shape and its chemical composition. Bone has also an ability to adapt itself to the externally imposed environment; thus bones of athletes are thicker and stronger, but bones of old people or people confined to bed become thinner because they are not required to transmit the high stresses of a young and active adult. The mechanism of this adaptation is still not well understood, but it was recognized already in 1892 by J. Wolff[1] and it has been known as Wolff's law from this time. This basic difference prevents us from making a consistent definition of bone as a material.

The main components which have an effect on mechanical properties of bone may be grouped in the following manner:

1. Crystalline mineral phase — hydroxyapatite.
2. Amorphous mineral phase — composite approximating hydroxyapatite.
3. Crystalline organic phase — collagen.
4. Amorphous organic phase — protein molecules in the form of gels and sols.
5. Liquids.

The existence of these various components suggests that bone may be considered a composite material, and then its function and its fracture characteristics may be better defined. Although some better understanding of bone by modeling it as a composite material has been achieved,[2,3] the properties of bone as a material were not specified because the relative amounts of the components and their distribution differs from one part of a bone to another.

The initial formation of bone starts from the deposition of a fibrous collagen to form the shape of the future bone. When collagen calcifies, the mineral phase is formed so intimately related to collagen, that fibrous structure of collagen is still visible in a fully matured bone.

II. FUNCTIONAL ADAPTATION OF BONES

Bones are structural components of the body and transmit predominantly compressive, torsional, and bending stresses. By offering high resistance to deformation, they also protect more vulnerable organs from injury. Both mechanical functions are performed with the minimum weight of the bony components and the maximum efficiency. It is realized only today in engineering that one of the best energy-absorbing mechanisms is a liquid-filled porous material.[4] Nature has designed the cranium to protect the brain from external injury in precisely this manner. Liquids in bone have, of course, another function recognized a long time ago; they serve as a transport medium for the nutrients which have to be supplied to cells, and for the waste products which have to be carried away. Soft tissue such as muscles, tendons, and ligaments transmit tensile stresses and apply compressive loads to bone.

Trabecular bone, which has a cellular, porous structure and is often referred to as cancellous or spongy bone, is designed primarily to absorb energy and transmit compressive stresses. The compressive strength of the trabecular bone is lower than the strength of the cortical bone which has a relatively solid structure, and hence this bone has to be thicker. The ends of long bones (epiphysis) which support the bearing surfaces of joints have to absorb energy, but also have to transmit compressive stress to the rest of the bone. Thus, it is quite logical that they are thicker and are made from a spongy trabecular bone rather than from a cortical bone. The fact that long bones are thinner in the middle (diaphysis) but still maintain the same strength has another reason. The distribution of weight-bearing material is such that the compressive stress remains the same throughout the whole length of the femur. However, the smaller size of bone in the middle reduces the strain in bending, thus increasing its ability to bend and reducing the probability of brittle fracture. With age, the proportion of solid bone decreases: the walls of long bones become thinner and cannot resist the same stresses. Nature compensates for this deficiency by increasing the outer diameter of bones, thus making them structurally stiffer. However, this also increases the amount of strain in the outer fibers on bending, and since bone is a strain-sensitive material, it becomes markedly more vulnerable to brittle fractures. It is also interesting to examine the cross-sectional shape of the femur. In engineering applications, the shape of the shaft depends on the type of stresses which the shaft is designed to resist. Thus, a cylindrical shape is the best for resisting torsional stresses, and a square shaft is optimal for resisting bending stress applied parallel to its sides. The shape of the shaft of the femur is triangular with well rounded corners, and the apex of the triangle points to the back of the body. This is the compromised shape for resistance to both torsional and bending stresses.

A. Microstructure of Cortical Bone

Figure 1 illustrates schematically the microcomponents of a cortical bone. It consists of four constituents:

1. periosteal bone having lamellar structure following the outer curvature of the cortex;
2. endosteal bone, also lamellar but with the radius of the lamellae corresponding to the inner radius of the cortex;
3. Haversian bone;
4. interstitial bone.

The whole length of a shaft of a long bone is surrounded by the periosteum on the outside and endosteum on the side of the marrow cavity. Periosteum and endosteum are made up of connective tissue and are packed with cells, having a very important biological rather than mechanical significance. The rest of the compact bone is also permeated with cells located in the lacunae or cavities, which have an approximately ellipsoidal shape. The shape and distribution of the lacunae, a discontinuity in the compact bone, has a mechanical significance when the resistance of bone to fracture is considered. Cells located in lacunae receive their supply of nutrients from the Haversian and Volkman's canals through the network of canaliculi.

The major structural component of the cortical bone is a Haversian bone, built up of osteons. Osteon is shown in Figure 2. Figure 2A is a photomicrograph taken with a reflected light microscope of a specimen polished for that purpose with metallurgical techniques. It shows that the Haversian canal, containing blood and lymphatic vessels, is surrounded by concentric lamellae. Figure 2B shows with transmitted light microscopy that the lacunae are also placed concentrically around the canal and connected radially with canaliculi.

Figure 2C is the same specimen viewed with polarized light. In this case it is possible to observe in more detail the lamellar structure of Haversian systems. The structure of concentric cylinders of osteons has been compared by Ascenzi and Bonucci[6] to cross-ply laminates, indicating that the fibrous structure of each lamella is oriented at a different angle to the long axis of an osteon. However, black and white coloring of lamellae may also indicate differences in composition or microstructural arrangement of organic and mineral phases within lamellae. Ascenzi has published more papers on the structure and properties of osteons than any other investigator. His views, however, on the fibrous arrangement in individual lamellae and lamellar interfaces have not been widely accepted and have generated controversy among many researchers. Since an osteon is a basic building block of a cortical bone, it is vital to know its structure. Figure 2D illustrates an osteon from the fractured surface of bone treated with ethylene diamine. This treatment removed completely an organic phase, illustrating that alternate lamellae are made of a continuous hydroxyapatite phase in poly-

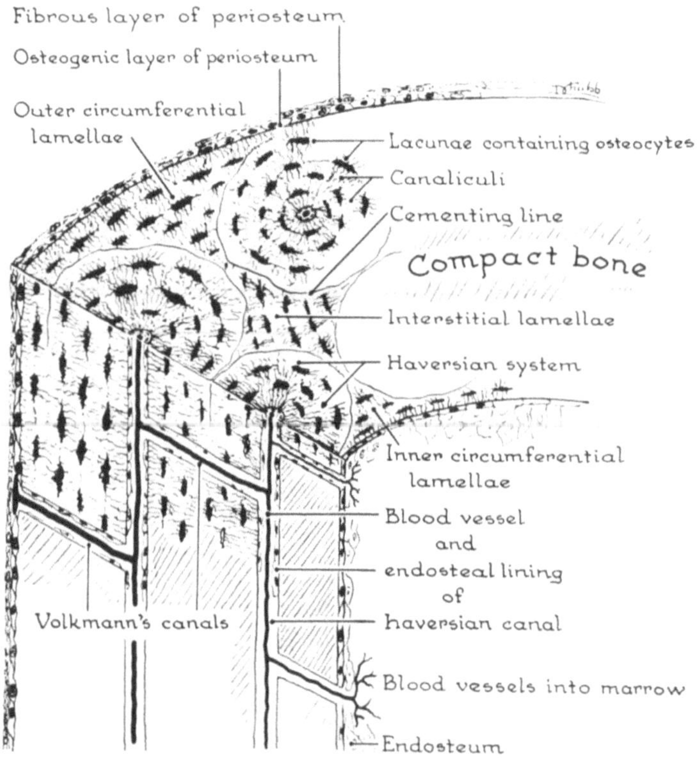

FIGURE 1. Schematic illustration of the microstructure of a cortical bone. (From Ham, A. W. and Cormack, D. H., *Histology*, 8th ed., Pitman, London, Chap. 15, 1977. With permission.)

crystalline form containing possibly some organic phase. The concentric cylinders of hydroxyapatite are separated from each other by empty spaces with some mineral cross-linking. It may be quite safely speculated from this photomicrograph that the empty spaces were filled predominantly with an organic phase or that the mineral present existed there as a discontinuous phase.

Clearly the above description of the lamellar microstructure of bone must have a great influence on the character, mechanisms, and mechanics of crack propagation through bone.

B. Fracture of Bone

Bone is a highly anisotropic material, on both microstructural and submicrostructural levels; consequently it is only logical that the energy to propagate a crack and the mechanisms of crack propagation depends on the spacial orientation of a test specimen. It was shown in 1966 by Bonfield and Li[7] that the energy absorbed during fracture is much higher in the longitudinal than in the transverse direction.

Energy absorption during fracture of bone has been measured by a number of investigators, and although the magnitude of the reported results differ, it is evident that energy to fracture decreases with age (see Table 1). It was also observed in this laboratory that even within the same age group, the scatter of the results is very large.

There is also a large number of researchers who have attempted to observe the mechanism of deformation in a cortical bone. Ascenzi and Bonucci[6] have tested single osteons in tension, and on the basis of stress-strain curves and microscopic examinations, concluded that characteristics of deformation depend on the orientation of fibers in an individual lamella. After the elastic limit of fibers has been reached, the interfibrilar cementing substance fails, and

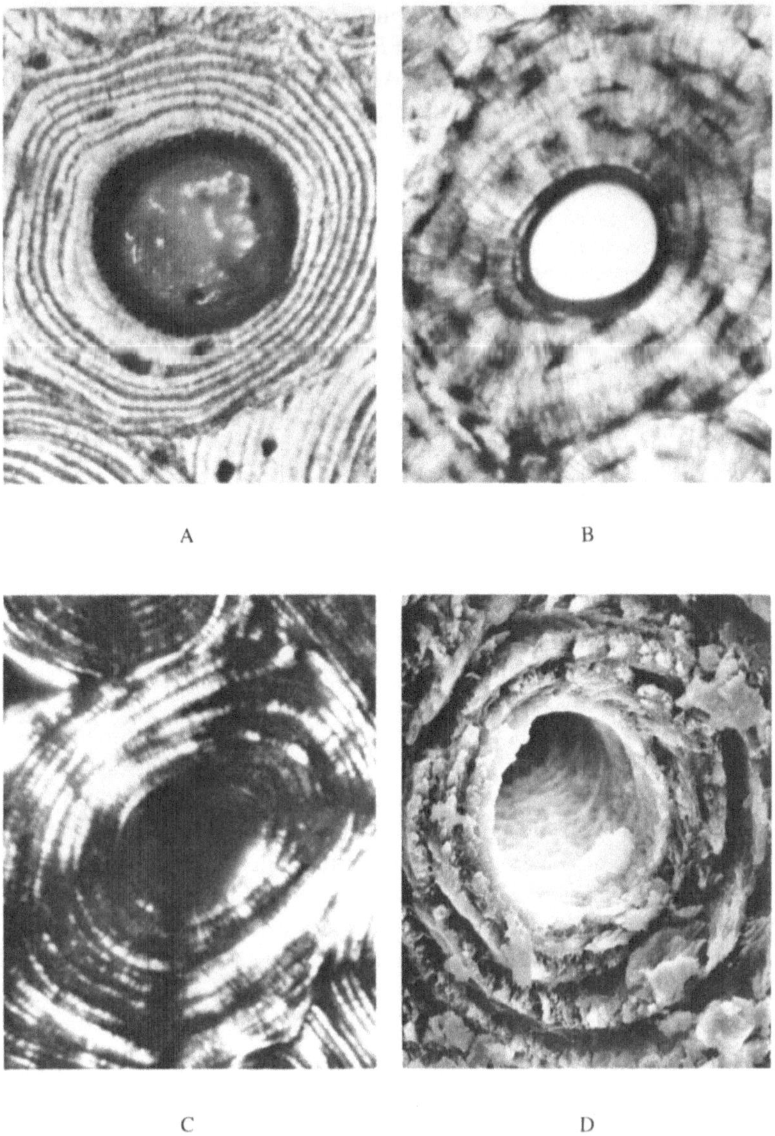

FIGURE 2. (A) Lamellar structure of an osteon (reflected light microscopy). (B) Osteon viewed with transmitted light. (C) Osteon viewed with polarized light. (D) Osteon from the fractured surface of bone, treated with ethylene diamine (SEM photomicrograph).

plastic deformation occurs, resulting in a series of microfractures. Although their photomicrographs show distinctly different lamellae (black and white), the lamellar pull-out has not been observed.

Most investigators have confirmed the previous findings in this study[8] by identifying two mechanisms of fracture: ductile — by the pull-out of osteons and fibers, and brittle — occurring mostly at higher strain rates, by the crack running indiscriminately across microconstituents.

Results almost identical with this study have been reported by Saha in shear[11] and in tensile tests.[17] Pre-cracking specimens before bending tests proved to be a very useful technique for Pope and Outwater[9] in order to obtain a controlled propagation of the crack. They have also attributed "plastic" deformation in bone to the inter-lamellar debonding

Table 1
ENERGY TO FRACTURE OF CORTICAL BONE

Type of bone	Joules/m^2	Ref.
Fresh bovine	59,000	8
Fresh bovine	58,650	10
Embalmed human	20,700	11
Fresh human adult	12,000—18,000	12
Fresh human children	16,000—22,000	
Fresh human 78—85 years	23,000	13
Fresh human 40—50 years	64,000	
Fresh human 10—40 years	22,000	14
(impact energy) 40—90 years	12,000	
Dry equine	1170—1340	15
Fresh bovine		16
crack velocity = 1.75×10^{-7}m sec^{-1}	764	
crack velocity = 23.5×10^{-7}m sec^{-1}	2125	
Fresh anthropoid		9
longitudinal	700	
transverse	1838	

mechanism. Currey and Butler[12] have also observed that energy absorption of the travelling crack is related to the roughness of the product surface. Comparisons of tensile tests with dynamic tests at strain rates of 133 sec^{-1} were made by Saha and Hayes.[18] A 34% increase in strength was observed in the dynamic tests, and high expenditure of energy on the initiation of a crack. Fractography also perfectly showed brittle failure in the dynamic tests.

It may be summarized that when fracture occurs parallel to the axes of osteons, at slow strain rates the crack propagates along weak interfaces which are found between concentric lamellae of osteons. Figure 3 shows such a crack propagation trying to go around osteons unless it is forced to go through one. It also has a biological significance, since a crack does not disrupt a blood supply to bone which is essential for its repair. In terms of energy absorption, the longer the path of a crack the larger the amount of energy absorbed.

When a crack is forced to propagate perpendicular to the axes of osteons, then at slow strain rates, the weak interlamellar interfaces also come into play and an osteon extends like a telescope. The final fracture of such extended osteons is shown in Figure 4. The mechanism of energy absorption is similar to one observed in the fiber-reinforced materials. When the interfaces between fibers and matrix are not too strong, a pull-out of fibers occurs and the friction of the pull-outs absorbs large portions of energy. Pull-outs of concentric lamellae of osteons act in the same manner, explaining partially strain rate sensitivity of a cortical bone. At high strain rates a crack propagates indiscriminately through the microconstituents absorbing much lower amounts of energy. Figure 5 shows fracture propagating at low strain rates from left to right and halfway through, changing into a rapidly propagating crack. The same observation is made by Behiri and Bonfield,[16] but the energy absorption occurred in the opposite manner (see Table 1).

C. Effect of Age on Fracture of Cortical Bone

It has been well established that the metabolic efficiency of organs and organisms declines with age. It is also a well-known fact that old people fracture bones much more easily than children. One of the reasons was mentioned in the introduction to this paper; the other clearly must be in the differences in the microstructure of a cortical bone.

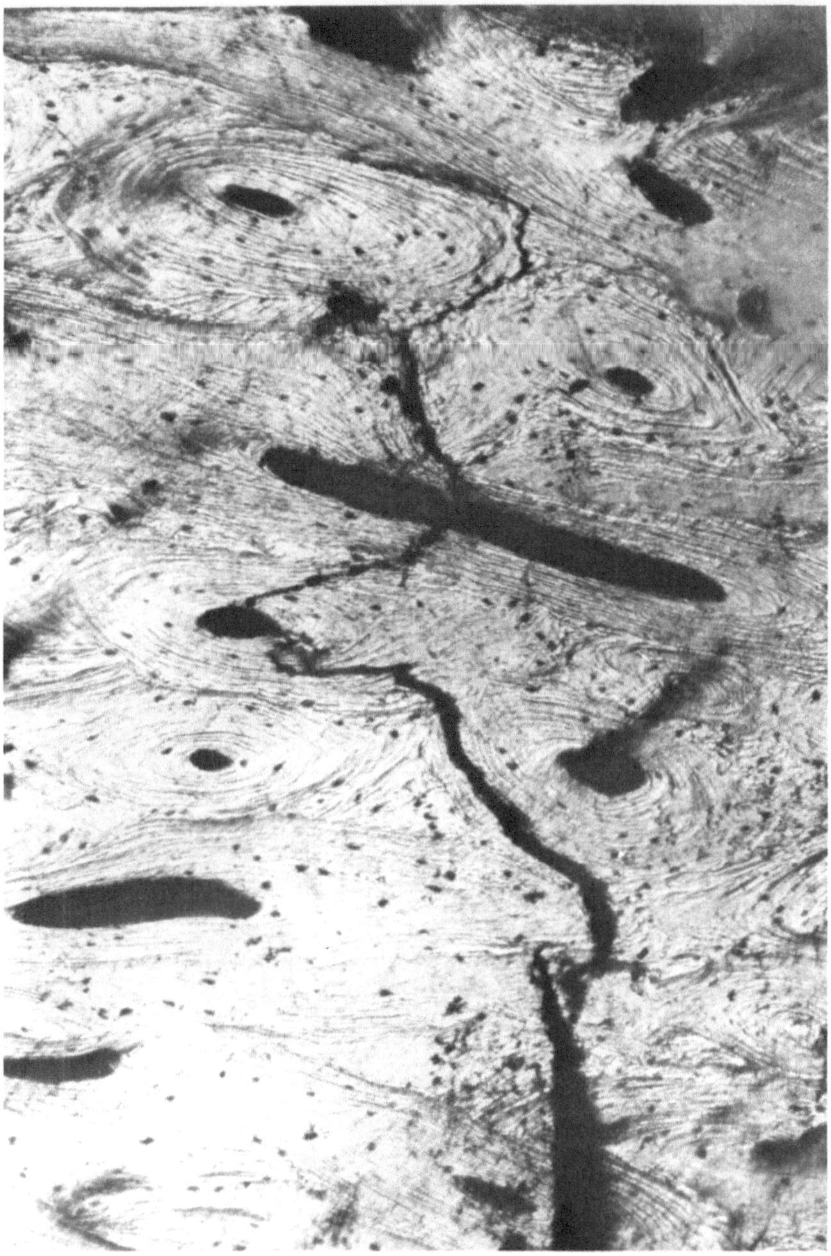

FIGURE 3. Crack propagation through a cortical bone at low strain rates and parallel to the axes of osteons.

It may be concluded from the reports of previous investigators that toughness of bone is related to the mechanism of crack propagation through the microstructural constituents of a cortical bone. The amount of energy absorbed is related to the osteon pull-out mechanism or production of a rough fracture surface. This, in turn, depends to a large extent on the rate of crack propagation.

In the author's 1976 study,[13] human bone from subjects of various ages obtained at post mortem or after amputations and preserved in deep freeze has been fractured at low strain rates. As a result, large differences in the energy absorption were noted (see Table 1), and

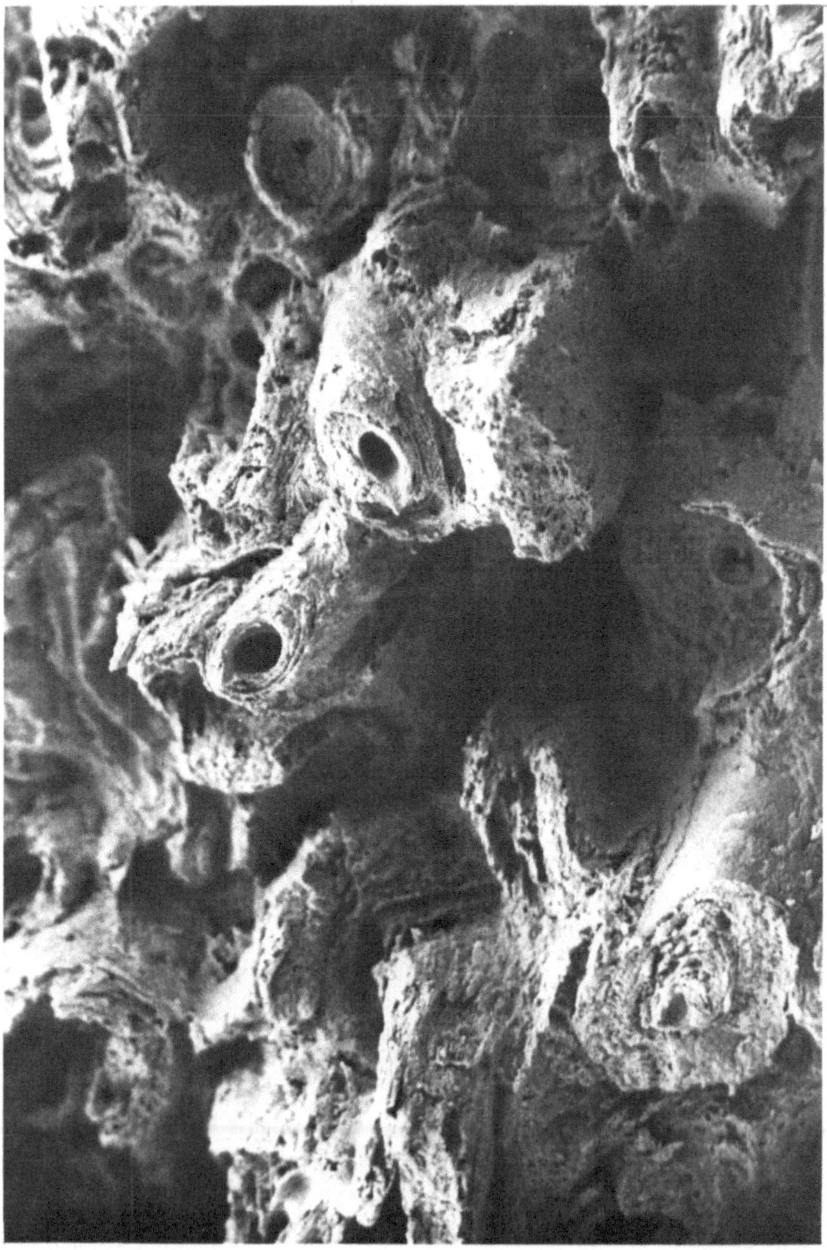

FIGURE 4. Fractured surface of a young human bone (23 years old) showing large portion of pulled-out osteons.

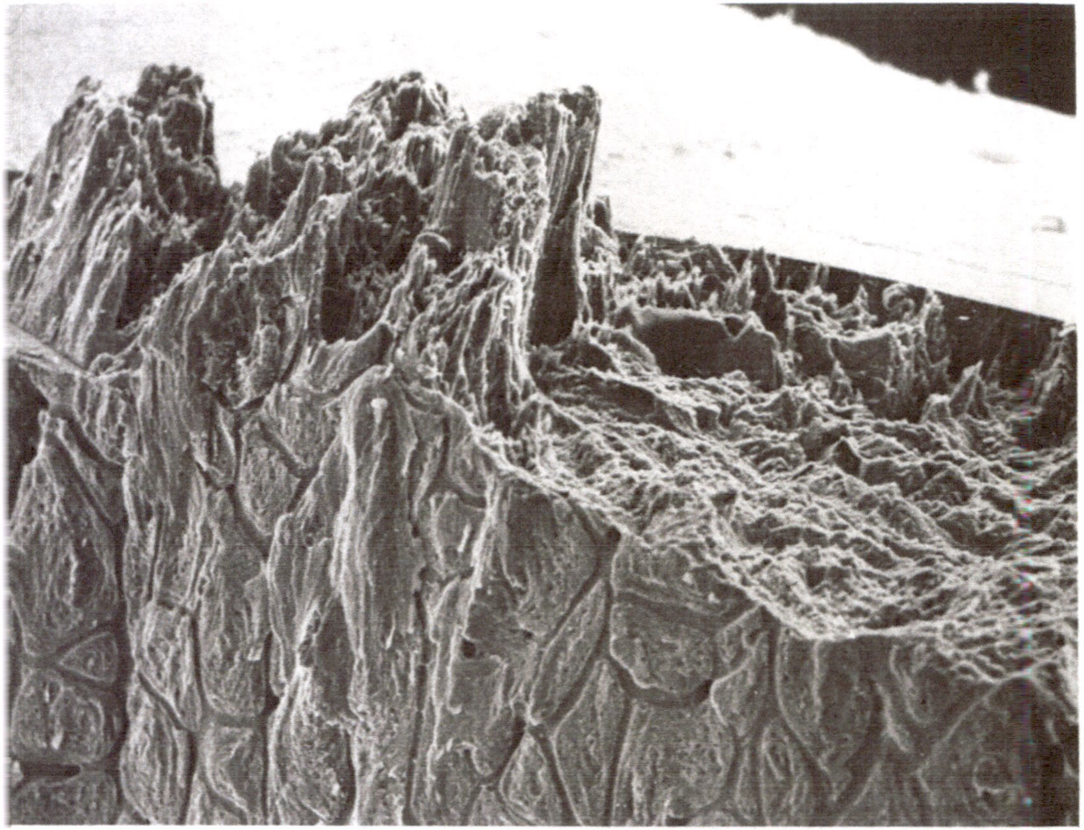

FIGURE 5. Slow propagating fracture (from left to right) developed halfway into a rapidly propagating crack.

the characteristics of the load deformation curves indicated the presence of either ductile or brittle material under test. The results obtained were so strikingly different that electron scanning micrography was used to determine the mechanism of crack propagation in brittle and in ductile specimens.

Fractography and measurements of energy required to propagate a crack confirmed earlier findings of other investigators.[12] Young bones absorbed more energy, and fracture was generally ductile. There were large portions of the fractured surface with pulled-out osteons clearly demonstrating the mechanism of energy absorption (Figure 4). Specimens from a human bone in the age range between 30 to 60 years did not produce any consistent results. Some fractures were ductile, some brittle, or a portion of the fractured surface would have pulled out osteons, and the part immediately adjacent to it would show a brittle fracture. Upon more accurate observation it was possible to distinguish osteons with entirely different morphology. Osteons contributing to the pull-out mechanisms demonstrated clear concentric lamellar structure with large gaps between concentric cylinders. The gaps were as large as in Figure 2D, where the organic phase has been removed. Osteons found in brittle bones did not show such distinctive concentric lamellae (Figure 6). There is some indication that such lamellae existed when an osteon was originally formed, but in time, radial calcification occurred bridging them into a solid mass. This may be the result reported first by Frost[19] and discussed by Currey[20] that in old bone when osteoblasts die in the lacunae, remodeling from the blood vessel does not always occur, but instead canaliculi connecting lacunae to the Haversian canal calcify. This radial calcification destroys the original lamellar structure of an osteon, instead of stimulating a remodeling process. It was also speculated by Currey that in spite of the fact that remodeling is necessary to keep the mineral of the bone available for exchange with the body fluids, bone still maintains its structural function and does not undergo a turnover without the live cells. Examples of such phenomena are observed in other mammals, and therefore, no matter how much we like to think of ourselves as being different, the possibility that the bone is dead before we are is a very real one.

Ductility of bone then does not depend so much on the age of bone as on the age of osteons. The largest ductility and energy absorption occurs in children's bones where there is not just the pull-out mechanism in operation, but also some ductility in partially pulled-out osteons which bend and result in a green stick fracture. The same mechanism may also exist in an adult bone as is shown in Figure 7. The fracture occurred during bending of bone at very low strain rates; the radius of bending was very small, and a pulled-out osteon was also bent showing the mode of a green stick fracture. The remarkable fact about this mode of deformation is that the bone specimen belonged to an 83-year-old man.

A hypothesis may be proposed about the aging process in bone. Young osteons are made up of concentric cylinders consisting of collagen and the continuous polycrystalline phase of hydroxyapatite. Between the lamellae is an aggregate of collagen and apatite which, during cyclic loading, transmits hydrostatic pressure like a liquid. This process, according to a model proposed earlier,[21] supplies nutrients to concentrically located bone cells. When the cells die, either remodeling occurs or radial calcification. Since the amount of remodeling decreases with age, old bones have more calcified osteons, and hence display a more brittle behavior.

D. Macrostructure of Trabecular Bone

Trabecular bone occurs in parts of the human skeleton. To illustrate its structure, a section through a human lumbar vertebra is shown in Figure 8. It should be noted that trabeculae have different shapes and different spatial orientation, presumably for the purpose of transmitting mechanical stresses more efficiently. On the right and left side, the curved shape of the vertebra is reinforced with relatively large plate-like trabeculae. In the center, the trabeculae of various shapes are arranged parallel to the axis of the vertebra taking care of

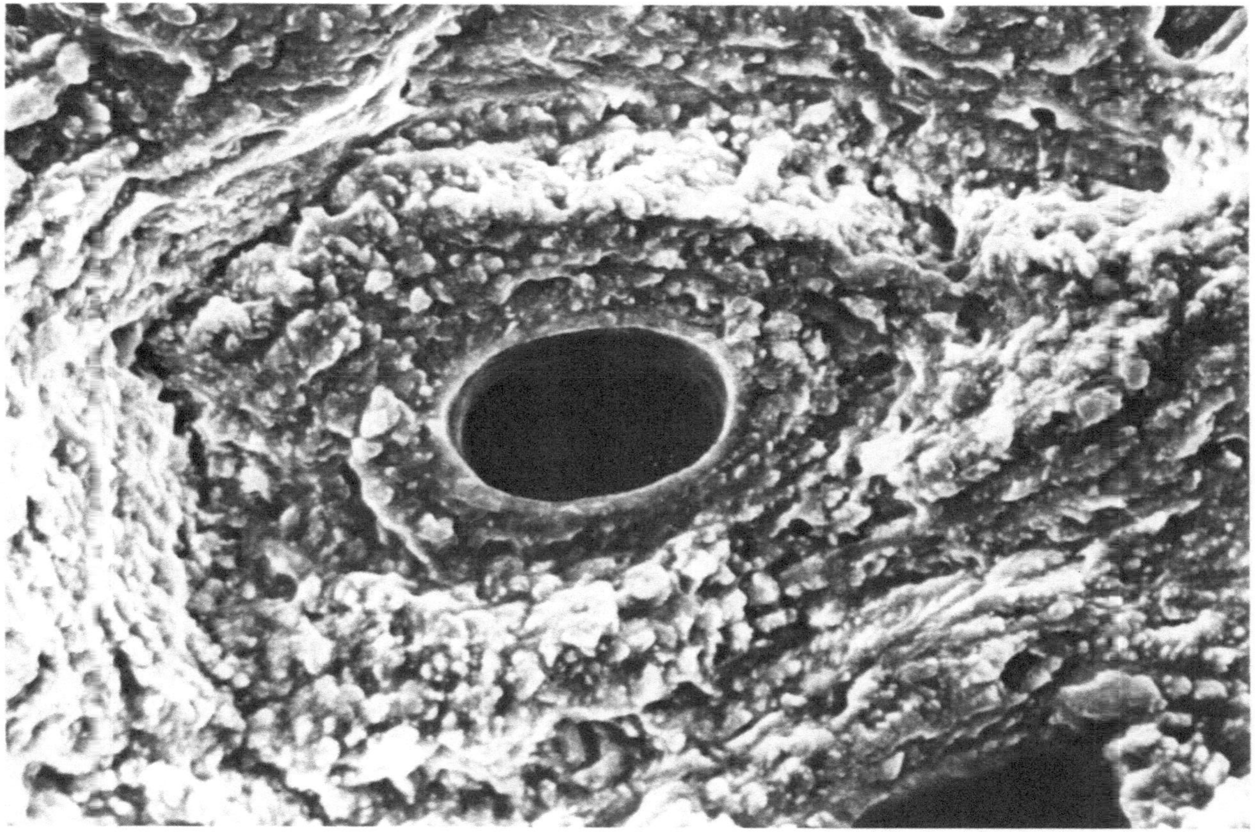

FIGURE 6. Structure of an old osteon indicating the previous existence of concentric rings which are presently bridged with the calcified tissue. Human bone (80 years of age).

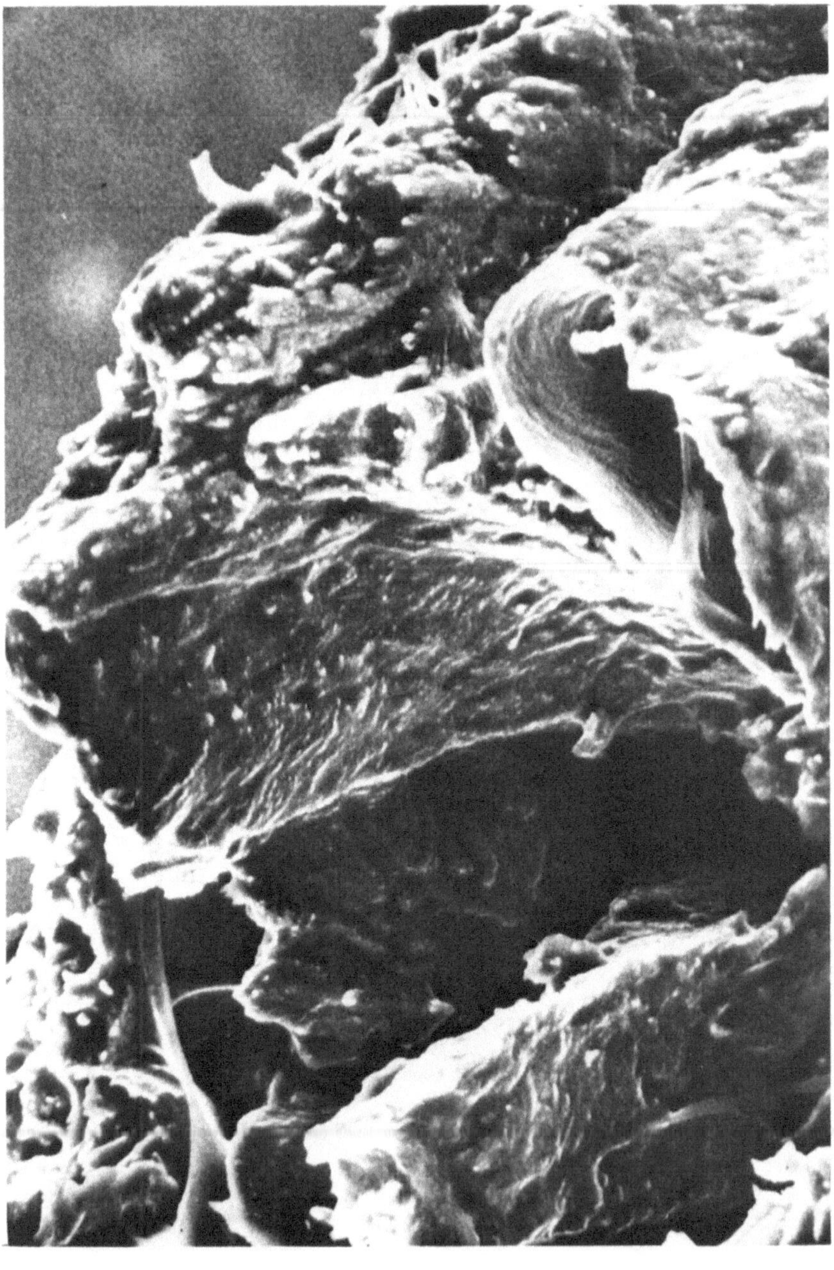

FIGURE 7. Young osteon in an 83 year old bone. The curvature of the pulled-out osteon caused by bending is similar to the "green stick" fracture.

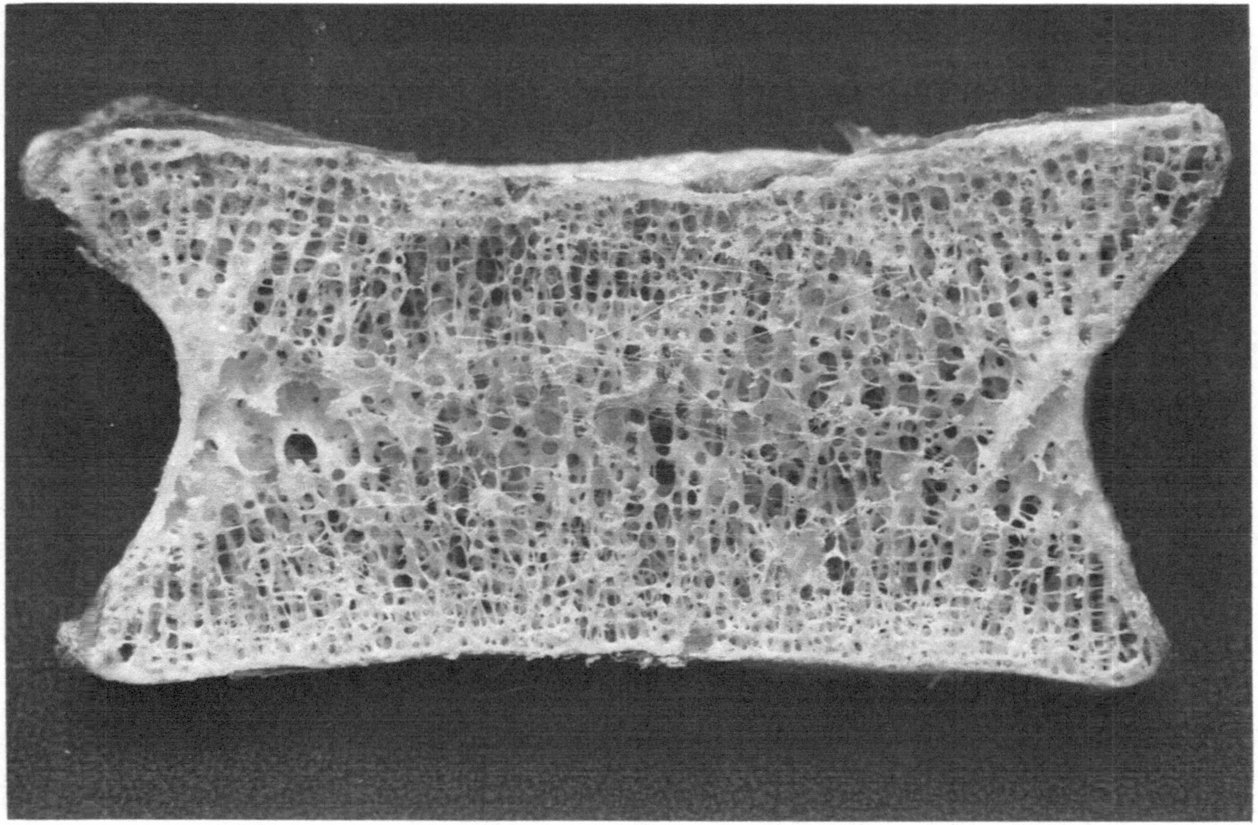

FIGURE 8. Typical transverse section of a human lumbar vertebra illustrating spacial arrangement of various types of trabeculae.

direct compressive stresses. The orientation changes in the four corners supporting curved portions of the outer shell. The outer shell is thicker than trabeculae and is made up of a Haversian bone having osteonic structure the same as described for a cortical bone. Most trabecular bone, however, does not have Haversian canals, and therefore lacunae with osteocytes have to rely on the supply of nutrients from the outside. For that reason, according to Ham,[5] the lacunae should not be positioned further away from the outside than 100 μm. Thus, the thickness of the most trabeculae is not greater than 200 μm.

The aspect ratio of trabeculae is generally small, i.e., the trabeculae are short to minimize the possibility of buckling under compressive stresses. This is further improved by the transverse struts supporting longer trabeculae. In addition, the cellular spaces, cancelli, are filled with the red and yellow marrow, which is highly viscous material and offers a very strong support at high strain rates. The marrow, which is permeated with blood vessels, is also a source of nutrients which travel through canaliculi to the lacunae located in the trabeculae.

E. Microstructure of Trabecular Bone

Figure 9 shows longitudinal and transverse section of a trabecula viewed under reflected light microscope and prepared by a metallurgical polishing technique. It clearly illustrates that trabeculae have also lamellar structure, and in a trabecula having rod-type shape the lamellae are concentric, very similar to the lamellae of osteons. In a plate-like trabecula the same structure becomes more elongated. At the junctions the trabeculae flair out with lamellar structure following the same shape. In order to be able to predict the behavior of trabeculae in fracture, it is necessary to learn more about the nature of trabecular lamellae. Figure 10 shows an SEM photomicrograph of such a lamella deproteinized with hydrozene. The similarity with an osteon from Figure 2D is striking. The solid mineral lamellae are separated from each other by hollow spaces. Here, it can be also speculated that hollow spaces were filled by a predominantly organic phase, possibly with the discontinuous mineral crystallites. Thus it may be safely postulated that the lamellae rich in organic phase are much weaker than the lamellae rich in mineral phase; therefore the deformation should occur by following the lamellar shape of weak interfaces.

Figure 11A illustrates the Gordon-Cook mechanism of weak interfaces, and Figure 11B shows the manifestation of this mechanism in the fracture of a trabecula. The fracture occurred in bending. The outer lamella was stressed in tension and fractured initiating a crack; then the crack propagated through several lamellae until the inner lamellae started to buckle in compression. Approximately at the neutral axis, delamination occurred opening a new crack at the weak interface running at the right angle with respect to the originally propagating fracture. In fractography of trabeculae such delamination along lamellae occurs very frequently, and it should be considered a typical deformation of a trabecular bone. Figure 12 is a three-dimensional view of the similar type of fracture. The outer lamella becomes separated, and the crack propagates longitudinally through the whole length of a trabecula.

III. SUMMARY AND CONCLUSIONS

Anisotropy of bone may be observed at various levels of magnification. At the "ultrastructure" the composite of hydroxyapatite crystals and collagen is so ultimately interconnected that separation of two phases during fracture implied by some of the investigators cannot really be documented. Collagen is naturally in the form of a fibrous structure. When calcification is completed the fibrous structure is still retained, but now it becomes a fibrous structure of a composite made up of a mineral and an organic phase. Thus, when the fractography shows a fibrous surface it does not necessarily mean that it is a pull-out of collagen fibers out of mineral or mineral crystals out of the organic phase; it is generally a

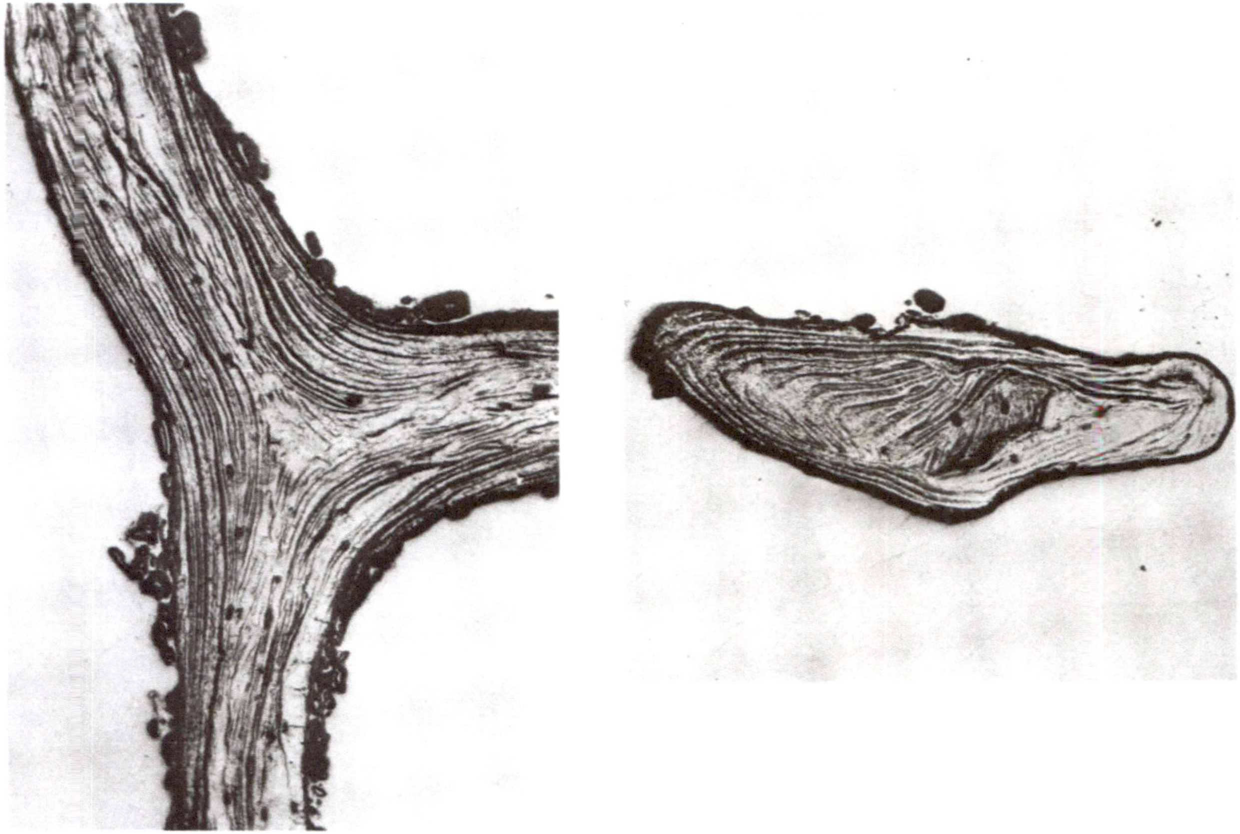

FIGURE 9. Longitudinal and transverse sections of a trabecula. *Specimen was polished and viewed with the reflected light.*

FIGURE 10. Lamellar structure of a portion of a trabecula deproteinized with hydrozene (SEM photomicrograph).

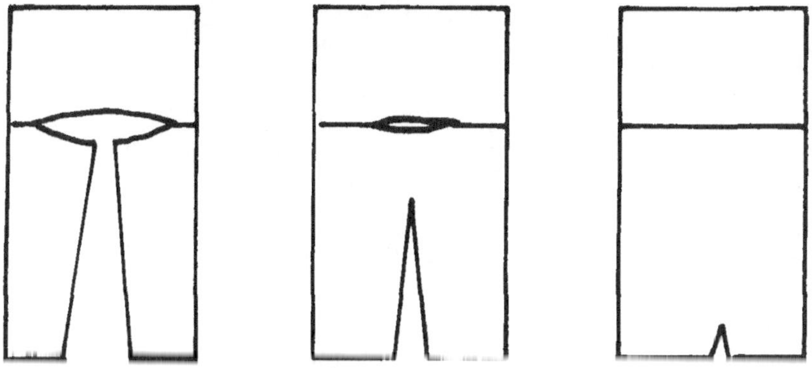

A

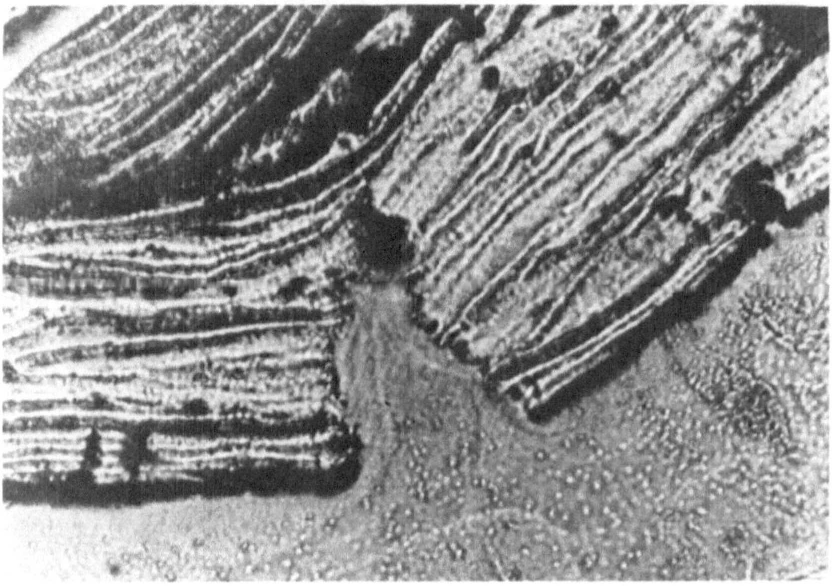

B

FIGURE 11. (A) Gordon-Cook mechanisms of propagation of a crack towards a weak interface. (B) Manifestation of this mechanism in the fractured trabecula.

fracture of a fibrous bone structure. The microconstituents of bone, however, arrange themselves in both cortical and trabecular bone in a lamellar structure in which some lamellae are rich in collagen and some rich in a mineral phase. Fracture and deformation occurs along weaker lamellae, increasing greatly the energy absorption of a deforming bone.

FIGURE 12. Fracture of a trabecula showing separation of the outer lamella and the longitudinal, interlamellar propagation of a crack (SEM photomicrograph).

REFERENCES

1. **Wolff, J.,** *Das Gesetz der Transformation der Knochen.* Hirschwald, Berlin, 1892.
2. **Katz, J. L.,** Hard tissue as a composite material — I, Bounds on the elastic behaviour. *J. Biomech.* 4, 455, 1971.
3. **Piekarski, K.,** Analysis of bone as a composite material. *Int. J. Eng. Sci.* 11, 557, 1973.
4. **Hilyard, N. C. and Kanakkanatt, S. V.,** Mechanical damping in liquid-filled foams. *J. Cell. Plast.* 87, 1976.
5. **Ham, A. W. and Cormack, D. H.,** *Histology,* 8th ed., Pitman, London, 5, 1977.
6. **Ascenzi, A. and Bonucci, E.,** Mechanical similarities between alternate osteons and cross-ply laminates. *J. Biomech.,* 9, 65, 1976.
7. **Bonfield, W. and Li, C. H.,** Deformation and fracture of bone. *J. Appl. Phys.* 37, No. 2, 869, 1966.
8. **Piekarski, K.,** Fracture of bone. *J. Appl. Phys.* 41, 215, 1970.
9. **Pope, M. H. and Outwater, J. D.,** The fracture characteristics of bone substance. *J. Biomech.* 5, 457, 1972.
10. **Melvin, J. W. and Evans, F. G.,** Crack propagation in bone. *Biomech. Symp.* American Society of Mechanical Engineers, 2, 87, 1973.
11. **Saha, S. and Hayes, W. C.,** Tensile impact properties of bone. American Society of Mechanical Engineers Symposium, AMD, 10, 125, 1975.
12. **Currey, J. D. and Butler, G.,** The mechanical properties of bone tissue in children. *J. Bone J. Surg.* (A), 57, 810, 1975.
13. **Piekarski, K.,** Unpublished results, 1976.
14. **Currey, J. D.,** Changes in the impact energy absorption of bone with age. *J. Biomech.* 12, 459, 1979.
15. **Alto, A. and Pope, M. H.,** On the fracture toughness of equine metacarpi. *J. Biomech.* 12, 415, 1979.
16. **Behiri, J. C. and Bonfield, W.,** Crack velocity dependence of longitudinal fracture in bone. *J. Mater. Sci.* 15, 1841, 1980.
17. **Saha, S.,** Application of electron fractography to bone fracture. Proc. 2nd New Eng. Bioeng. Conf. 1975.
18. **Saha, S. and Hayes, W. C.,** Relations between tensile impact properties and microstructure of compact bone. *J. Biomech.* 9, 243, G. B. 1976.
19. **Frost, H. M.,** In vivo osteocyte death. *J. Bone J. Surg.,* 42A, 138, 1960.
20. **Currey, J. D.,** Stress concentrations in bone. *J. Micros. Sci.,* 103, Part 1, 111, 1962.
21. **Piekarski, K. and Munro, M.,** Transport mechanism operating between blood supply and osteocytes in long bones. *Nature* 269, No. 5623, 80, 1977.

Chapter 8

STRUCTURAL ASPECTS OF TENDONS AND LIGAMENTS

Ian W. Forster

TABLE OF CONTENTS

I. HISTORY AND INTRODUCTION

Historically tendons were not studied effectively until Von Haller[61] in the 18th century showed tendons to be insensitive. Although Galen had described tendons in the 2nd century, he confused them with mixtures of tendon and nerve, and clear distinction between these two structures was not made until Von Haller's experiment. The first person to operate on tendons was an Arabian physician called Aricenna in the 11th century. Naturally, repairs of these structures before Von Haller were very disappointing, and many authors, e.g., Guillemeau in 1598 and Falcon in 1649,[38] strongly advocated against intervention after injury. Guillemeau even stated "When nerves and tendons are stitched on account of the pricks made by the needle, pain, swelling, inflammation and convulsions ensue and often death on account of the sympathy of the principle and origin which is the brain". Later the opposite view was held. Hunter, in 1767, looking at the repair process compared it with that of bone, Viering[135] later showing that cells arose from both peritendinous and intratendinous tissue. He also showed that reorganization of the tendon required function. Bichat in 1802 recognized a fibrous system that extends everywhere, and "it provided for its basis with a fibre of peculiar nature, hard rather elastic miscible. Hardly admitting of contractibility sometimes in juxta position and assembled parallel to each other in tendons and ligaments; at other times intersected in various sections, as in membranes, the capsules, the fibrous sheaths but everywhere the same".[14]

Bichat's conclusions were based on the boiling and drying of tendons. He also experimented to determine the tensile strength of tendon and pointed out that it was largely inelastic, giving away completely when the sudden effort was too great. He was obviously the forerunner of collagen research.

Mason and Sheardon[88] (in 1932) reinvestigated Von Haller's findings, stressing the importance of the blood supply to the tendon, and stated that success of any repair depended on a delicate balance between immobilization to allow strong repair and early movement to prevent adhesion.

Exactly the same problem is examined in publications today: controversy still exists on whether or not to repair Achilles tendons and even over the actual function of the knee ligaments.

J. L. Petit (1674 — 1750)[104] pointed out that surgery was not necessary for healing of Achilles tendon, and healing at the correct tendon length was achieved by plantar flexion of the foot and knee flexion using an "Apparatus".

In the same book Cooper described Alexander Monro Primus, the great Edinburgh Surgeon (1697 — 1769), rupturing his own Achilles tendon. He immediately pushed his foot downwards fully and pushed the calf muscles towards the heel to achieve apposition until help came. His final effective appliance consisted of a slipper with a belt attached to the heel which was also attached to another belt around the calf. The two could be tightened and the calf and heel brought closer together as required. After 2 weeks the foot was gradually brought upwards, until he could walk with a raised heel at 8 weeks. He was free from such restraint at 5 months. The factor which decided when Munro passed from one phase of remobilization to another was the size and apparent density of the tissue at the site of injury. This is not necessarily a good indication of healing!

It is interesting that controversy still exists over whether to repair ruptured Achilles tendons or treat them conservatively.[27] Difficulties have also arisen over the anatomical details of, for instance, the anterior cruciate ligament, and over the interpretation of crude in-vitro experiments and the extrapolation of the results into the clinical situation. This has led one authority to suggest that isolated cruciate ligament rupture does not occur and others to argue exactly the opposite.[73,77,98]

II. AIMS OF THIS CHAPTER

Bearing this controversy in mind, this chapter will describe the function of both ligaments and tendons, relating this where possible to structure. I will include biomechanical tests that have been applied to them, relating the results to the effects of disease and injury.

Tendons and ligaments are considered under the same heading, because histologically they have the same structure, being aggregation of dense connective tissue. However, their functions are very different.

III. DEFINITION AND ANATOMY

A tendon is the means by which a muscle contraction can be turned into force about a joint. The tendon therefore begins at its junction with the muscle fibers and ends on insertion to the relevant bone beyond the joint around which it acts. This simple definition does not take into account that there are basically two types of tendon: those with short excursion — such as Achilles tendon, and those with a longer distance over which to slide — such as the flexor tendons of the fingers, usually close to bone.

The first will repair with adhesion formation; but this will not greatly affect function. The others have a surrounding synovial sheath, and when adhesion occurs between the sheath and tendon, function is lost whatever the ultimate strength of the tendon itself.

Ligaments however, are more uniform. These are much shorter and act as stabilizers both inside and about the joint from the bone proximal to the distal end of the joint. Some ligaments are thickenings within the fibrous joint capsule, whereas others, such as the cruciate ligaments are entities in themselves. When considering any ligament we should think not only of the bone-ligament bone complex itself, but of the other structures — particularly muscles and other ligaments — acting simultaneously, which may add to its effect.

Both tendons and ligaments may cross several joints, e.g., the flexor tendons of the fingers which start at the lower forearm and continue to the distal phalanx of the finger; and the ligamentum nuchae which arises from spinous processes of the cervical spine and is attached to the occiput at the back of the skull.

The gross anatomy of each ligament and tendon is outlined in anatomical textbooks. The best histological description is found in Ham and Cormack[62] and Bloom and Fawcett.[16]

IV. CONNECTIVE TISSUE HISTOLOGY

Histological textbooks describe tendons and ligaments as being derived from connective tissue. Connective tissue is one of the four basic tissues of the body and, as its name suggests, connects other tissues together. There are three types of connective tissues:

1. Loose connective tissue — soft tissue layer with a large amount of intercellular ground substance and some elasticity determined by collagen elastic and reticular fibers. This loose connective tissue is found throughout the body covering organs, especially vessels and nerves. It is also a major component of the subcutaneous tissue which we will contrast with the other forms of connective tissue later.
2. Blood and other hematoopoietic (blood-forming) tissue, such as marrow.
3. Strong connective tissue, which is further subdivided into (a) dense connective tissue, e.g., tendons, (b) cartilage, and (c) bone. Dense ordinary tissue consists mostly of collagen fibers with occasional elastic ones. Since these fibers are extra-cellular and "non-living", few blood vessels are required for the cells within the connective tissue which produce the proteoglycan ground substance surrounding the fibers. Collagen fibers are stained pink with hematoxylin and eosin but are green on staining with

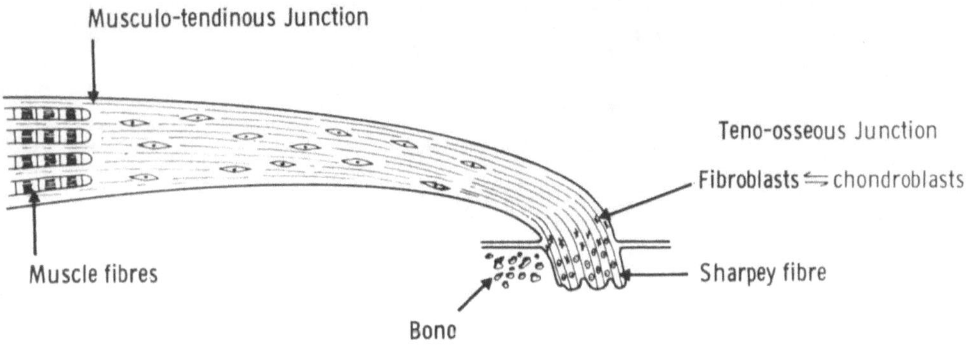

FIGURE 1. Schematic diagram of a tendon and its attachments.

Goldner's trichrome. They are partly birefringent, so their presence is enhanced by the polarizing microscope. This is the characteristic appearance of the dense connective tissues which comprises tendons and ligaments.

V. ORGANIZATION OF CONNECTIVE TISSUE

Dense ordinary connective tissue may have regular or irregular orientation of its fibers. That with regular fibers has great tensile strength and is well able to resist stretching. The irregular orientation is much less able to do this, and is normally seen only in muscle aponeurosis where fibrous tissue around one muscle blends with another so that there is a multidirectional "pull" or tension of the aponeurosis.

VI. HISTOLOGY OF LIGAMENTS

The dense connective tissue of the fibrous capsule surrounding joints is organized into several ligaments which are cord-like thickenings of it. In addition some ligaments, though separated from the capsule, are nevertheless derived from it, or have sometimes developed from the tendons of muscles which have become redundant during natural selection. At the attachments of the ligaments to bone, their structure becomes fibrocartilaginous, the collagen fibers being intermingled with intercellular ground substance and the fibroblasts resembling chronodrocytes. The fibers insert into the cortex of the bone. These gradual changes are shown by a blue appearance of the fiber and the cells, using specific stains for cartilage (toluidine blue).

VII. HISTOLOGY OF TENDONS

Tendons arise from a muscle at its musculo-tendinous junction (Figure 1). Muscle tissue has collagen fibers around and among its fibers. These come together and amalgamate into the dense connective tissue of tendon, so that at the junction, tongues of muscle fibers appear to be inserted into the tendon. The sarcolemmae of the muscle fibers are also directly attached to tendon connective tissue. At the other end of the tendon, collagen fibers insert through the periosteum around the bone into the cortex itself. The insertion is similar to that of ligaments, and the actual fibers are called Sharpey's fibers.

VIII. SYNOVIAL SHEATH

Tendons passing next to bone and requiring a long excursion over several joints are

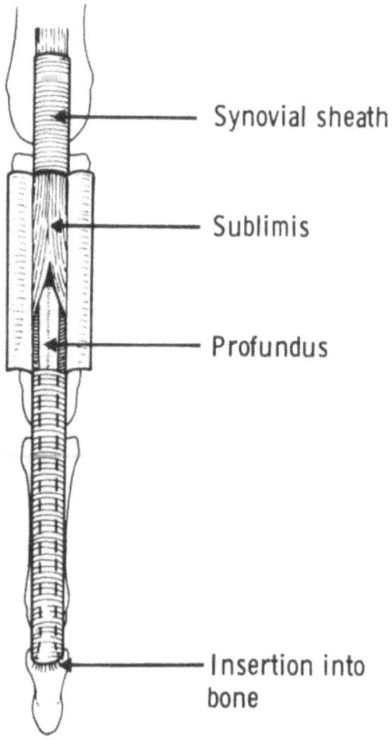

FIGURE 2. Arrangement of digital flexor tendons within the synovial sheath.

surrounded by a tendon sheath (Figure 2). This sheath consists of two layers, an outer fibrous and inner synovial layer. Synovial fluid is secreted into the sheath, and it is possible that tendons derive nutrition from this fluid.[87] In the finger flexors, the fibrous layer is condensed over the middle of each phalanx to prevent bowstringing of the tendon at full flexion of each joint and thereby improving the tendon's efficiency. These condensations are called "pulleys".

IX. BLOOD SUPPLY

Both ligaments and tendons have a blood supply, though in the normal state both are relatively acellular. This supply mostly comes from networks of blood vessels around the structures, but where synovial sheaths are present the networks are supplied by individual vessels on long slender loose connective tissue strands travelling across the fluid space within the sheath (Figure 3). They are thus liable to rupture, and even stretching of them can devascularize the tendon supplied by that vessel.

X. NERVE ENDINGS AND SENSORY ADAPTATION

Nerve fibers are present in dense connective tissue. Both ligaments and tendons are well supplied with sensory nerve endings called Golgi tendon organs. The functions of the tendon organs and their relation to muscle contraction were elucidated in a paper by Stephens et al.[123] Electrical activity was observed after both active contraction and passive stretching. The greater response following active muscle contraction was thought to be due to summation of the stimulation of many Golgi tendon organs rather than increased firing of a single organ. Proske and Gregory[108] confirmed that branching of the mechanoreceptors may extend the

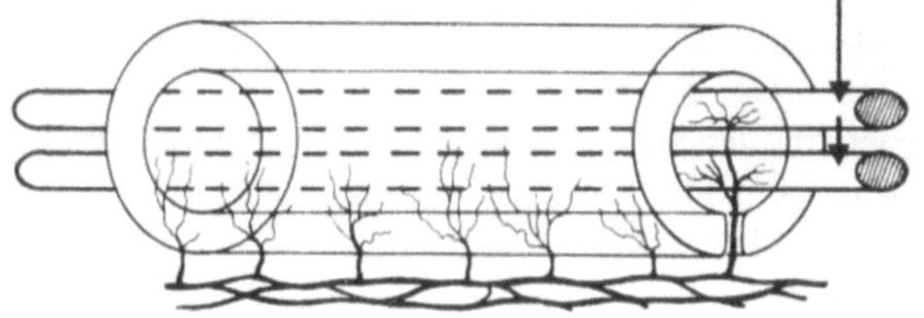

FIGURE 3. Blood supply to the flexor tendons passing through the tube of the synovial sheath and crossing the space to enter the tendons.

sensitivity range of the receptors. This was confirmed by Binder et al.,[15] observing single motor units firing in soleus tendon of the cat. Gregory et al.[56] pointed out the possibility of a late super-normal period in the recovery of excitability in tendon organs. What this means is that after a period of stimulation and discharge of the Golgi tendon organ, the discharge could be restarted by a stimulus which would not normally have done so; this indicates that the sensory organs are hyperexcitable immediately after their activity stops. Such a mechanism prevents repeated smaller stretchings of the tendon injuring it by summating their effects.

Nerve plexuses and sensory endings were described by Rozovskaya and Yankovskaya[114] in the human flexor digitorum profundus and flexor digitorum superficialis tendons and in their sheaths.

Kennedy et al.[77] have described ramifying myelinated nerve fibers in the anterior cruciate ligaments by silver impregnation techniques. As yet it is unknown whether sensory endings are present after reconstruction of joint ligaments by transfer or regeneration. We are undertaking experiments to study this aspect of tendon function, as it may have a bearing on why reconstructions using transferred tissue seem to lengthen at a later stage.

When muscle was stretched beyond its optimal length (Stauffer and Stephens[122]) the possible contractile forces are reduced. In contrast, the passive forces within the muscle increase, so that the overall force is unaltered and stimulation to the Golgi tendon organs is unchanged. These sensory endings therefore fail to fire. The muscle is protected in this instance by its own muscle stretch receptors which are in parallel with its fibers. The same situation does not occur in ligaments, because they do not have receptors acting in opposition to one another.

XI. DEVELOPMENT AND REPAIR

The adult dense connective tissue contrasts sharply with the mesenchyme in the embryo. In a developing tendon, dense bundles of fibroblasts proliferate, secreting more and more collagen and becoming gradually less cellular. The cells are oriented from the beginning in the direction of the applied stress. Tendon sheaths develop where the tendon itself may rest on bone.

This development is closely mirrored during repair. Fibroblasts proliferate from either the sheath or the loose connective tissue surrounding the tendon or ligament. These orient along the lines of stress and initially have a good capillary blood supply. From these cells, collagen is formed and gradually as more and more is laid, down the number of cells decreases as does the blood supply. The fibroblasts themselves arise from mesenchymal cells, the pre-existing fibrocytes taking no part in repair.

Interestingly Ippolito et al.,[68] when examining the Achilles tendon of New Zealand rabbits, found the tendon cells in longitudinal parallel rows. Fine filaments of 50 to 70° were found in the cortical area of the cytoplasm, which reacted strongly with anti-smooth muscle antibodies under immunofluorescence. He suggested that tendons can contract; moreover, when muscle suddenly contracts its effect may be modulated by the tendon relaxing. No other workers had described this effect in normal tendons. Nevertheless "myofibroblasts" have been described in regenerating tendons,[106] healing wounds,[49] and in diseased tissue, e.g., Dupuytren's contracture in the hands.[50]

The cells and their contractile elements appear to be a feature of developing tissue and not of mature tissue. However, in Dupuytren's disease this developing state appears to last for years in a relatively aged population!

XII. THE COLLAGEN FIBER

How does a ligament withstand the high stresses placed upon it? To answer this question we need to examine both the fine structure of the collagen fiber and its biochemistry.

A. Fine Structure

Collagen fibers under the electron microscope are seen to consist of fibrils of 0.3 to 0.5 μm diameter with an axial periodicity of 640 Å. Thus the fiber appears as a long cylindrical structure with alternate light and dark bands (Figure 4). Naturally the number of fibrils in each fiber is variable. The collagen molecule itself is some 2800 Å long and 15 Å thick. By staggering a bundle of these molecules on one another within a fibril by $^1/_4$ of their length, an axial periodicity of 640 Å is created. Ham and Cormack[62] suggest that a collagen molecule is a triple helix of 3 amino acid chains. In more recent work, it has been shown that up to 5 chains may be involved in the helix.[44,105] Indeed not only is there axial staggering, but Parry and Craig,[103] using x-ray diffraction, have shown lateral staggering of 80 Å, which corresponds to a repeating unit of 4 microfibils. This is seen especially in tendon and contributes to the special shape of each fiber. The microfibrils in each unit are related to one another in a 4_3 screw arrangement.

Petkov[105] by decomposing and washing off the cementing matrix of ground substance, found the natural collagen consisted of spiral fibrils or filaments whose diameter was 60 to 110 Å, with an angle of inclination of 85 to 90 degrees. Each filament consisted of 3 to 5 subfilaments each of 30 to 45 Å. It was much more difficult to split the interfibrillar connection than the fibrils themselves, and this important aspect will be discussed later. Confirmation was provided by Lillie et al.,[83] who displayed the helical structure of collagen by causing acutely angled spiral clefts in its surface, using solvents (Figures 4 and 5); and by Belton et al.,[12] who produced cleavage along the line of least resistance in a freeze etch study of collagen.

B. Biochemistry of Collagen

Collagen is a protein and therefore is composed of a sequence of amino acids. It contains an unusually high percentage of proline and glycine and is unique in that most of the proline and lysine is hydroxylated. Cells capable of producing collagen must therefore contain the enzymes which hydroxylate proline. This amino acid is capable of forming crosslinks and binding amino acids together. Rojkind,[112] in a review of collagen biosynthesis, described collagens as a heterogenous class of extracellular proteins characterized by a unique amino acid composition. They contain 30% glycine, 20% proline and hydroxyproline, and a variable amount of hydroxylysine and carbohydrate.

Collagens according to Rojkind consist of 3 left-handed chains of 100,000 daltons (chains), which are assembled into a right-handed superhelix of 300,000 daltons. Each polypeptide chain has an helical extension of 15 amino acids at both its carboxy and amino acid terminal ends. These extensions contain the lysine and hydroxylysine residues, which can be enzymically attached to aldehydes which are the precursors of crosslinks (Figure 6).

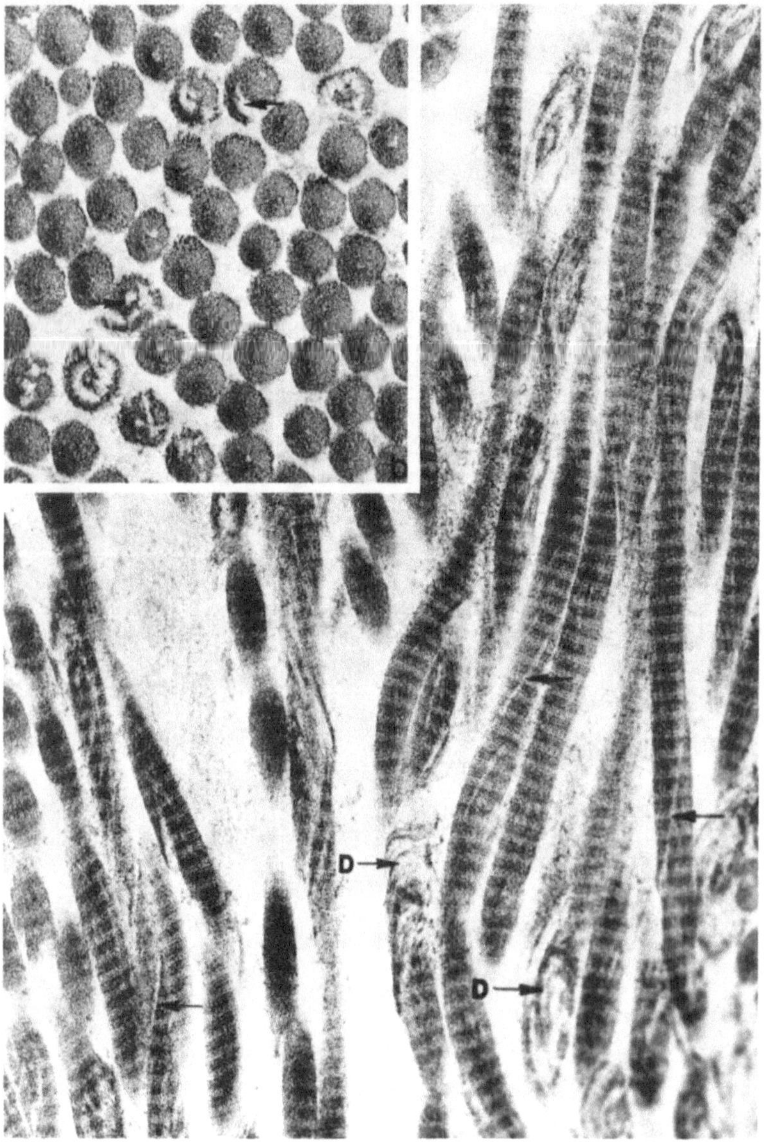

FIGURE 4. The Collagen Fiber. (A) The helical arrangement of fibrils is shown following solvent action producing spiral clefts. (B) Fine structure. (From Lillie, J. H., MacCallum, D. K., Scaletta, L. J., and Occhino, J. C., Collagen Structure: Evidence for a helical organization of the collagen fibril. *J. Ultrastruct. Res.*, 58, 134, 1977. With permission.)

These crosslinks may be

1. *Intramolecular* between two aldehyde residues of neighboring chains in a single collagen molecule.
2. *Intermolecular* between the aldehyde residue of one chain and the ϵ amino groups of lysine or hydroxylysine of a second triple helix. Exact fiber alignment (i.e., $^1/_4$ overlap) is necessary for this to happen.

Collagen chains also include glucose and galactose residues attached to the hydroxyl

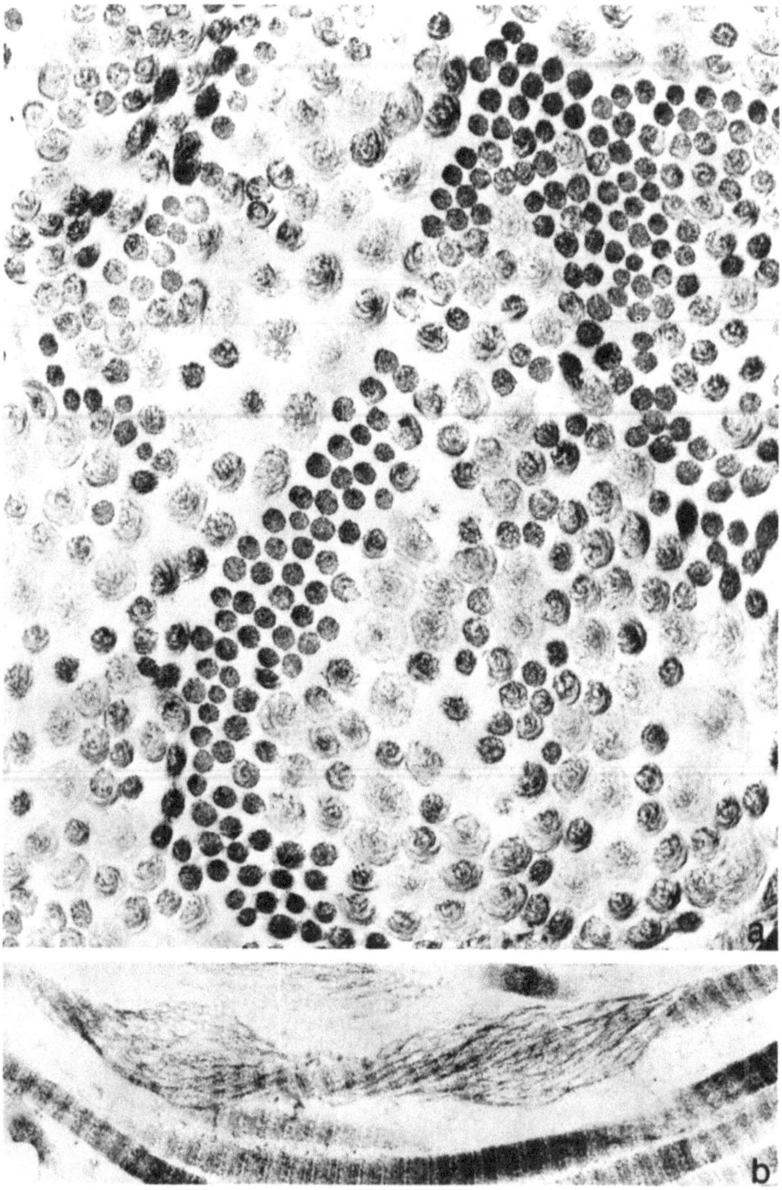

FIGURE 5. (A) Helical arrangement within collagen fibers. (B) Fine structure. (From Lillie, J. H., MacCallum, D. K., Scaletta, L. J., and Occhino, J. C., Collagen Structure: Evidence for a helical organization of the collagen fibril, *J. Ultrastruct. Res.*, 58, 134, 1977. With permission.)

terminal of hydroxylysine through a glycosidic linkage. This is mainly in the disaccharide form, glucosylogalactosylhydroxylysine. These linkages will be discussed in greater detail later.

There are several distinct types of collagen isolated from various tissues. We will consider in detail only collagen I and III, as these are the only ones involved in normal tendons and ligaments and in their repair.

Type I contains two identical chains and one different. It is the major collagen in bones and tendons and contains glycine 10% alanine, 10% proline, 10% 4 hydroxyproline 0.7 hydroxylysine, and 1% carbohydrate.

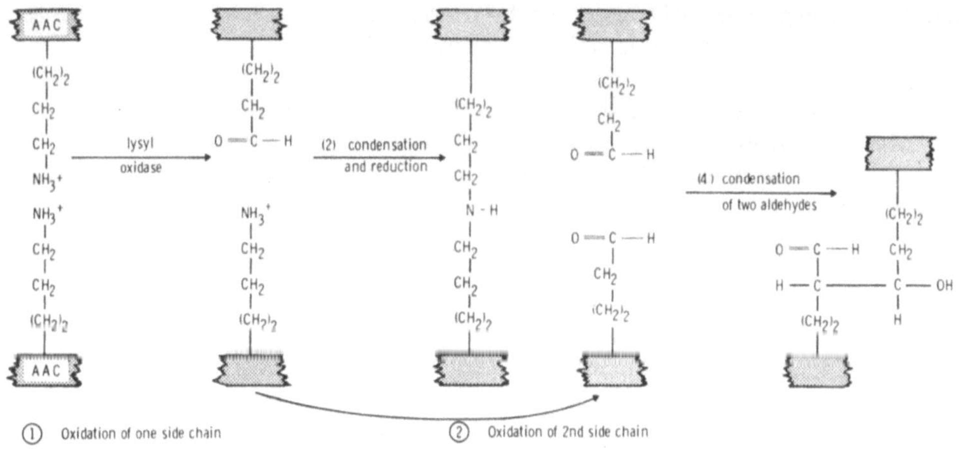

FIGURE 6. Cross-linking in collagen.

This collagen, when dissolved by proteolysis using pepsin, lacks non-helical extensions and precipitates out of solution from 0.05 Tris HCl buffer, pH 7.4 with 2.5 NaCl. It corresponds to the heavy collagen bundles that stain blue with trichrome and light brown with reticular stain.

Type II is the only collagen present in elastic and hyaline cartilage.

Type III is present in most tissues containing Type I collagen except for bone and tendon. It contains more glycine and more hydroxyproline. As it also contains two residues of cystein per chain and it is able to form disulfide crosslinks in addition to the aldehyde linkages. It precipitates out of solution from Tris buffer with 1.7 *M* NaCl and corresponds to the fine argentophilic fibers seen microscopically.

Type IV is the collagen of basement membrane. It also has been isolated from skin, placenta, liver, spleen, and pancreas.

Type V is produced only by two embryonic tumors and is similar to Type I, except that the 3 chains are identical.

XIII. COLLAGEN TYPING

If strength is to be ascribed to one Type of collagen rather than another, it is important to be able to distinguish between the relative amounts of Types I and III in any tissue under consideration.

Heathcote and Al-Alawi[63] have used a technique involving partial hydrolysis of the dermis, using hydrochloric acid in a nitrogen atmosphere. A two-dimensional thin-layer chromatography of the fluid under investigation is then carried out and compared with a sample from a normal person. The differences are subtle, and it is felt by this author that a better and more obvious difference is shown in the method used by Coombs et al.[33] The fresh collagen from an Achilles tendon is broken down to its constituent polypeptides, using cyanogen bromide. The typing of these polypeptides is performed by simple electrolysis on polyacrylamide gel in the presence of sodium dodecyl sulfate. Using this method, the differences between collagen types are much more obvious. This method is also quantitive.

XIV. COLLAGEN BIOSYNTHESIS

By creating collagen fibrils in vitro it is possible to study collagen synthesis and organization and the factors which may affect this. Gelman et al.[51] started with collagen molecules free from aggregate and with intact non-helical terminals. This is essential for collagen biosynthesis. He found that at least three steps were involved:

1. **Initiation**, which has a temperature-dependent intermediate.
2. **Linear growth**, which is temperature independent.
3. **Lateral associated of filaments**, which is also temperature dependent and takes time to develop.

The intermediate in (1) was thought to be a microfibril which is perhaps a five-fold D helical. Step 2 could occur by the NH_2 group of one end reacting with the COOH of another. Temperature independence would be helped by electrostatic stabilization. Step 3 — covalent crosslinking might occur if preformed aldehydes were present.

All these changes are chemical and show the intimate interrelationship between chemical reaction and structure. I shall later show how intimate is the relationship between chemical stability and mechanical strength. The formation of collagen involves both intra- and extra-cellular phases. These events are summarized by Veis and Brownell[133] and by Rojkind.[112]

A. Cellular Events

The first event to occur is the synthesis of procollagen which corresponds to a single chain consisting of an initial 50,000 dalton non-collagenous terminal, a middle organized section with a glycine molecule at every third residue, and with a common sequence of Gly—Pro—Pro (or Hydroxypro). The sequence is determined by procollagen messenger — RNA. This extends for 1,000 residues followed by a further non-collagenous terminal.

These events occur initially within the membrane-bound ribosomes, and the formed proteins are then transported through the membrane to the cisternae of the endoplasmic reticulum.

Trelsted and Hayashi,[130] using embryonic forelimbs of mice labeled with H_3 Proline and then fixed at intervals, showed increased uptake in the Golgi bodies. The Golgi bodies are specific organelles within a cell visible with an electron microscope and not to be confused with Golgi tendon organs. Their function is to alter the composition of proteins synthesized elsewhere. Using the electron microscope "Segment long spacing", filaments can be seen within the vacuoles of the Golgi bodies. These are approximately 300 μm long and are equivalent to procollagens. The cross striation was not seen within bundles at this stage. This confirms the earlier findings of Frederickson et al.[46] and, in isolated tendon cells, of Kao et al.[75]

Typical collagen fibrils were later found in deep recesses of the cell surface and Trelsted considered that ultimate length of the fibril would be able to be controlled by these recesses if there was a "sensation of tension" in the cell wall. The lack of striation would suggest that crosslinking occurs in the extracellular space.

Both Veis and Brownell[133] and Rojkind[112] thought that hydroxylation of proline occurs as an intracellular event while the amino acid chain is still elongating. This process requires enzymes which can be specifically inhibited by α,α dipyrindyl. After this, the rate of collagen synthesis was drastically reduced.[119,133]

The enzymes are also inhibited by corticosteroids or by a deficiency of ascorbic acid[133] though Kao[74] does not agree that this is necessarily the cause. The hydroxylation of proline appears to have a stabilizing influence on the triple helix fibril, although no specific crosslinks has yet been ascribed to it.

Cellular adhesion is thought to be a possible mechanism involved in polymerization of collagen — Goldberg,[53] Hahn and Yamada,[59] and Linsenmeyer et al.[84] have all suggested this; possibly the glycoprotein fibronectin is the agent involved.

B. Extracellular Reactions

The extracellular reactions are those of crosslinking. Naturally, because of the possibilities of chemical reorganization, great interest has been taken in this phase.

The initial reaction is a cleavage of the procollagen at both its terminals by several steps using different enzymes;[139] probably the amino end is detached first. At least five enzymes are involved in the post transitional stage.[110] After this is the formation of intra- and extra-chain crosslinks.[111] Most of the work in this field has been carried out using a well-described technique of borohydride reduction. This has been able to break and characterize both types of crosslinks.

Sodium borohydride may divide intermolecular crosslinks, and sodium cyano-borohydride the intra-molecular crosslinks. The enzymes involved in the formation of those crosslinks have been partially characterized; lysyloxidase which oxides specific lysyl and hydroxylysyl groups is well-known. Fujii et al.,[48] however, suggested that lysyl links are not so important as was previously thought, and suggested that histidine residues may have an important role. Others[34] suggest that only modifying lysine and hydroxylysine affects the formation of stable crosslinks.

The problem, eloquently pointed out by Viidik,[138] is that many crosslinks occur in vitro which do not exist in vivo. Another is that preparation and preservation of tissue may well affect the results.[65]

Nevertheless, it seems likely that the aldehydes of lysine and hydroxylysine, i.e., allysine and hydroxyallyl-lysine, respectively, appear to react together to form crosslinks. Aldelydes of two adjacent residues form intra-molecular bonds which are very stable in themselves. These bonds are temperature stable, and most chain-splitting techniques divide the inter- rather than the intra-molecular links (Figure 6).

The inter-molecular link is by a reaction between the aldehyde and the amino groups of lysine or hydroxylysine. This is rather unstable and has to be restabilized by a reduction or arrangement of the double bond or by the formation of further more complex crosslinking using histidine[48] and/or glucose and galactose residues in collagen.[99]

Papers have described the roles of, for instance, Cis — trans isomerization of peptide bonds[7] and even dihydro—dihydroxylysino—nor—leucine.[18] Bachinger[7] suggested that three interchain disulfide bridges form a nucleus for the triple helix, which in itself is temperature dependent and that the two kinetic phases of helix or coil are interchangeable and that isomerization into the superstructure of a coil occurs with this type of bond. These bonds take time to develop, and a changing proportion of them is a feature of both aging and resistance to stress.

The structural stages involved in formation of a tendon are summarized in Figure 7. The tropocollagen becomes amalgamated into a micro- and then a subfibril. It is only at the next stage of fibril formation that 1/4 staggering and axial periodicity is seen; crimping (vi) appears only at the fascicle stage.

Bruchner et al.[21] compared the amino terminals of Types I and III collagen, with the intention of looking at differences in bonding. He showed that there was a rapid triple helix-coil transformation in Type III, because of an increased number of peptides allowing disulfide bonds. This is a fully reversible temperature-dependent reaction, and mismatching of the chains does not occur. Nevertheless, although the bonds are the same as in Type I, the proportion is different, and the ultimate strength of this collagen is not as high as Type I. The ground substance surrounding the collagen is a glycosaminoglycan (GAG). This is the new terminology, the previous name being mucopolysaccharide. Naturally this has given rise to some confusion. Using an immunological technique involving antisera to GAG, Anderson et al.[3,4] have isolated this glycoprotein from bovine Achilles tendon and, using this technique, investigated the effect of tendon glycoprotein on fibril formation from collagen solution. They measured the spectrophotometric absorbance of a collagen gel at 400 nm

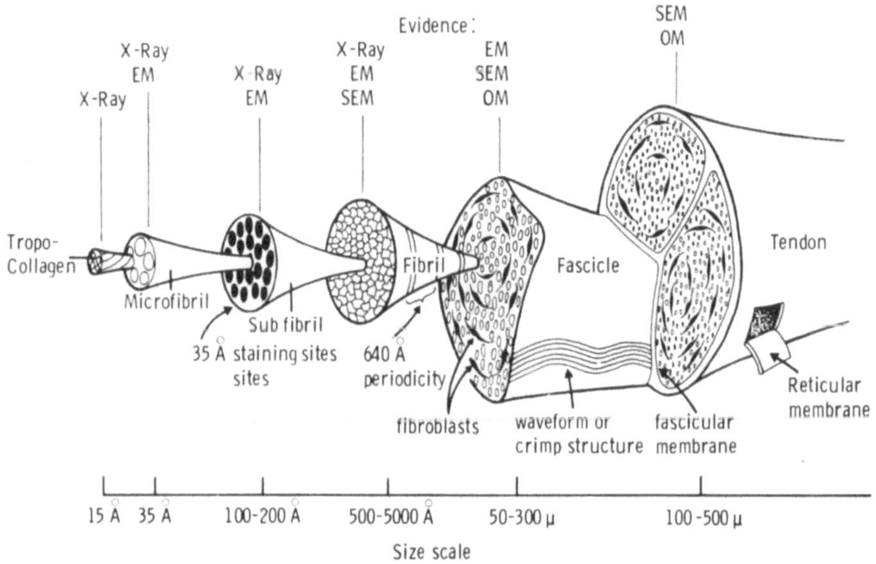

FIGURE 7. Tendon hierarchy. (From Kastelic, J., Galeski, A., and Baer, E., The multicomposite structure of tendons, *Connect. Tissue Res.*, 6, 11, 1978. With permission.)

(400 Å), this being a function of fibril formation. At 37° they noted that GAG decreased the final 400 Å and retarded fibril formation, with those formed also being thinner and more sensitive to changes in ionic strength, pH, and temperature.[3,4] Glycoprotein was found to be associated with collagen in the gel phase only when higher concentrations were used, with the resulting weak binding. Therefore smaller concentrations are essential for collagen fibril formation.

XV. STRESS RESISTANCE OF COLLAGEN FIBERS

The strength of each individual fiber is more or less the same. The major differences between fiber bundles is in their alignment — thus skin with its criss-cross network is relatively elastic and weak, whereas tendon with its longitudinal tight packing of fibers is inelastic and extremely strong. The resistance to stretching nevertheless takes place at the crosslinks. Several investigators have attempted to study tension at this point. Mohanaradhakrishnan et al.[94] showed that in dumbbell sections of skin, differences in strength depended on fibril orientation rather than fibril type.

There is still controversy whether the fiber is crimped in vivo or whether this is an experimental artifact. Most people accept the former now, and Kastelic et al.[76] have shown how it occurs. Chatterjee et al.[29] confirmed this to be a normal finding by making a surface model of the individual fiber from a mold. Indeed Frazer and MacRae[44] had to decrimp the rat tail tendon they were investigating (by placing it under tension) in order to perform x-ray diffraction studies. This crimping is a means of altering tendon length, but at present it is not considered to be important.

XVI. ULTRASTRUCTURAL DEFORMATION OF COLLAGEN

Barenberg[9] used compression-extension of extracted fibers of rat tail tendon on a variety of adhesive substrates. The non-polar regions of collagen were the most deformable, and these are the most organized areas within the fiber. No difference of deformation charac-

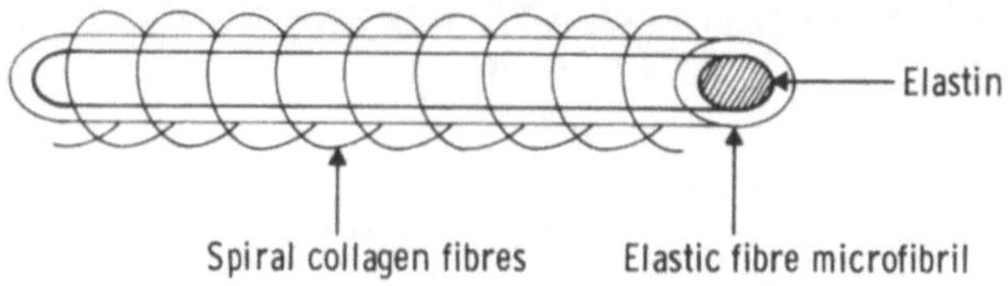

FIGURE 8. Schematic diagram of an elastic fiber.

teristics occurred with age despite the increase in insoluble collagen from 20 to 88%. Reconstituted fibrils reacted in the same way. This suggests that the intramolecular crosslinks give way. It is at variance with the fact that on standard stress testing intermolecular links are reversible but intramolecular links are not. An important reason why there is this variance, is the number and different types of non-physiological tests used to investigate crosslinking and that in vitro experiments can only ever offer a possible pointer towards in vivo physiology.

Other papers using heat[2] as a means of developing tension in the skin showed different results, tension being maintained in neonatal tissue to 95°C while adult tissue gave way at 75°C. They related the changes to increased amounts of heat labile aldimine links. Their views agree with those of Viidik.[138]

Histological staining of collagen fibers with Masson's trichrome both relaxed and under tension showed a color change with increased tension. Flint[39] suggested that this might be due to the increased availability of amino binding sites under tension.

Bell[11] has attempted to stimulate wound contraction by incubating fibroblasts of different proliferation potentials with prepared lattices of collagen. After 1 week the area of the lattice occupied 1/28 of the original area. He offers no biochemical explanation of this phenomenon.

XVII. ELASTIC FIBERS AND LIGAMENTS

Elastic fibers are formed in a similar way to the collagen ones. They consist of two distinct portions: elastin which is insoluble and the elastic fiber microfibril.[113] This microfibril at first surrounds the formed elastic fibers and is also present in the interstices of them, so that it occupies 10% of the whole fiber (Figure 8). The elastic fiber microfibril is a glycoprotein but lacks the hydroxylation of lysine and proline seen in collagen. It has a larger proportion of cystein and consequently of disulfide bonds. The fibrils develop in clefts in the cell wall, which allows control of polymerization, aggregation, and direction of fibers. During development the elastic fiber microfibrils appear first, elastin gradually increasingly until at maturity this is by far the most predominant part of the elastic fiber.

Serafini-Fracassini,[117] discussing the ultrastructures of elastic ligaments, suggested that the major central part of elastic ligaments consisted of longitudinal elastic fibers surrounded by a sheath of spirally oriented collagen and with an intervening layer of microfibrils. The proportion of elastin to collagen was 4:1.

Because of this construction, one would expect a biphasic stress stain curve — initially from the elastin and later the collagen.

However, he pointed out that in elastic ligaments the elastic fibers themselves stop short of the bone insertion, and here longitudinal collagen enters the bone, producing Sharpey's fibers as in the more usual purely collagen ligaments. Under the initial load therefore, stress is transmitted through the terminal longitudinal collagen fibers. The spiral sheaths elongate and narrow, by which time the interfacial frictional forces are sufficient to transfer any elongation to the elastin under physiological stress. This appears to be biphasic, the later modulus being a function of both the elastin and the tensile modulus of collagen.

Not only are elastic fibers stretchable in this way. One writer has suggested that intermolecular crosslinks may be elongated so that secondary stabilizing units may slip reversibly by 2 to 3%, this being increased by low angle spacing of them.[19]

XVIII. FUNCTION AND ULTIMATE STRENGTH OF TENDONS AND LIGAMENTS

Both tendons and ligaments are strong, and injury when it occurs usually does so at the junction with bone or with muscle, unless there is some inherent defect within the tendon itself. One can therefore test either the ligament itself or the bone ligament bone unit as a whole. Tests carried out in vitro are not physiological because the direction of force is simplified to allow analysis. In contrast, in life injury is caused by many different multidirectional forces; moreover the experimental force is usually a gradually rising force until the tissue yields, whereas injury is caused by a sudden severe force. Nevertheless comparison can be made with the living situation, although the relevance of any tests carried out must be questioned.

As far as a tendon is concerned, function is not only as a result of ultimate strength but may also require that the tendon moves adequately within the enclosing sheath. The function of a ligament clearly depends both on the forces which it is expected to resist and on their direction. Clinically, compensation for both these aspects may occur. Thus, a patient with minor knee instability who puts little load on his knee will not have any symptoms; and one who does not require fine movements of his fingers in order to work will cope adequately with a stiff finger after adhesions within the tendon sheath have reduced his tendon movement. One of the disappointing aspects of the knee reconstruction is that the transferred "ligaments" always lengthen with time; this may be due either to maturation and therefore removal of contractile elements or to lack of sensory nerve endings which would limit stretching.

Measuring the ultimate strength of a ligament does not measure the effects of division of the ligament; thus two types of functional testing are usually carried out: (1) stress-strain extension of the ligament, and (2) laxity after division of the ligament.

Both these studies have a bearing on the clinical features of injury, but both necessarily give confusing results. The vast majority of work has concerned knee ligaments for two reasons; first, they commonly present difficult clinical problems; and second, the ligaments are large enough to be easily testable. The majority of our attention will therefore be toward the knee joint. In connection with tendons we shall deal with the digital flexor tendons. This is because they present problems of the same magnitude because of adhesions around them after injury.

XIX. ANATOMY OF KNEE LIGAMENTS

The knee has four basic stabilizing ligaments between the femur and tibia: two collateral ligaments over the medial and lateral aspects of the joint; and two cruciate ligaments across the center of the joint. The cruciate ligaments cross as their name suggests, but not nearly as much as the older anatomical text books have described. The role and anatomy of these ligaments is still under discussion (Figure 9).

According to Warwick and Williams,[142] the anterior cruciate arises from the lateral femoral condyle within the intercondylar notch and inserts into the anterior medial aspect of the upper tibia over a wide area. The fibers are not solely longitudinal; some are also arranged spirally. Girgis et al.[52] and Kennedy et al.[77] have pointed out that the origin of the ligament in the femoral condyle is not uniform in that the anterior fibers are attached at a slightly lower level than the posterior, giving a trapezoid appearance to the ligament rather than an oblong (Figure 10). Further, these writers recognize two parts to the ligament — the anteromedial and larger posterolateral parts. Clearly, as the knee bends, some of these fibers will be taut and become lax, others loose and become taut. These functions can be ascribed to the posterolateral and anteromedial bands, respectively. As this is so simple a concept, it is difficult to see why such confusion has existed and still exists over the exact functions

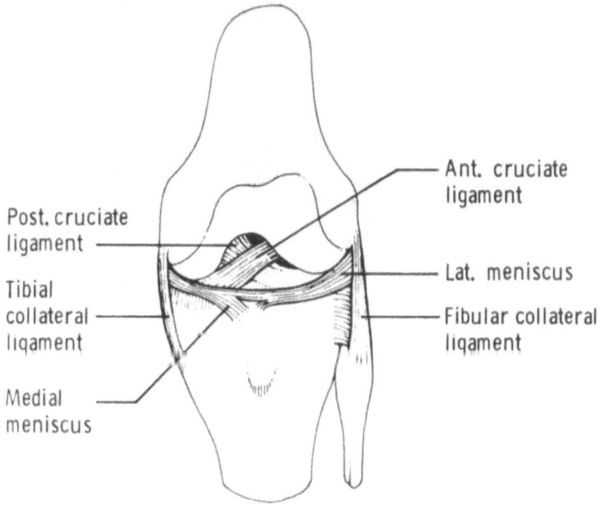

FIGURE 9. Knee ligaments viewed from the front.

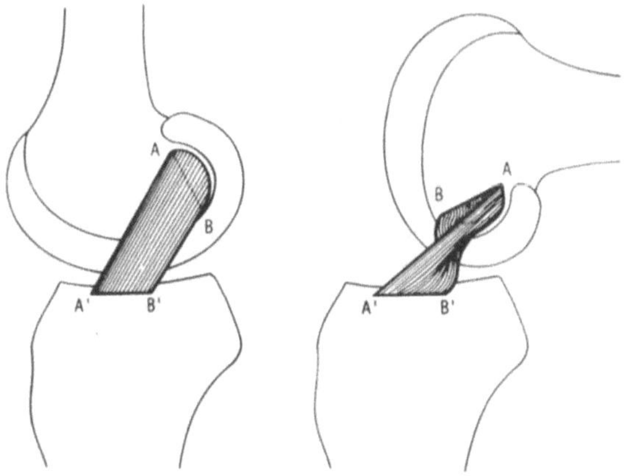

FIGURE 10. Schematic drawing representing changes in the shape and tension of the anterior cruciate components in extension and flexion. In flexion, lengthening of small medial band (A-A′) and shortening the bulk of the ligament (B-B′). (From Girgis, F., Marshall, J. L., and Al Monajem, A. R. S., The cruciate ligaments of the knee joint, *Clin. Orthop. Relat. Res.*, 106, 216, 1975. With permission.)

of the ligament and over its relative tightness with respect to the stage of knee flexion.[20,21,57,60,81,115]

The reason is simple; it is because all of these papers are the results of observations of knee flexion in vitro, often with other ligaments divided, the other stabilizing structures of the knee such as muscles being removed. It has been shown[20] that even the order in which ligaments are divided in these experiments has a profound bearing on the results. Thus fewer experiments are now carried out on dissected knees. More are performed on the intact knee, using radiopaque markers and x-rays, or measuring strain within the intact ligament.[78,131] Clearly the length of the ligament is crucial to its function. Various methods of measurement of ligament length and strain, using three-dimensional analysis with recording of coordinates, have been attempted in the isolated knee;[131] in such experiments torque was equated with

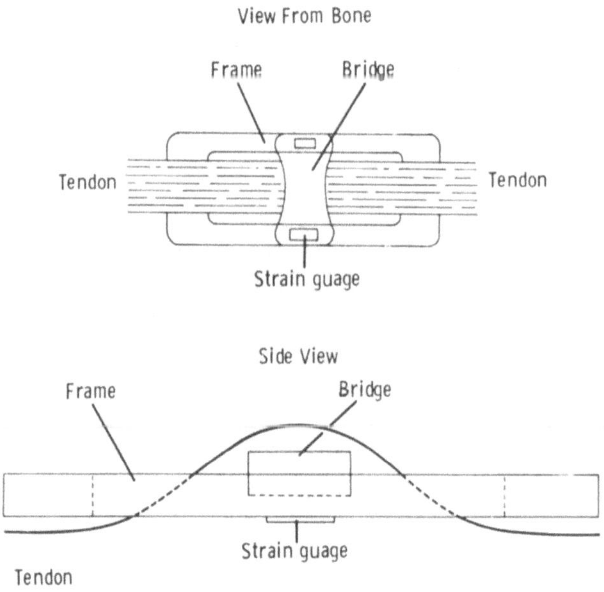

FIGURE 11. Schematic of buckle transducer. (From Barnes, G. R. G. and Pinder, D. N., In vivo tendon tension and bone strain measurement and correlation. *J. Biomech.*, Pergamon Press, 7, 35, 1974. With permission.)

ligament stretch. This force was applied until the individual ligaments failed, the other ligaments having been previously divided. Lew and Lewis[82] approached the problem in a similar way, suggesting the use of coordinates for radiographs in order to use a highly complex mathematical method in vivo; unfortunately, in both these techniques the placing of the coordinates is difficult and rather arbitrary, and this leads to at least 10% inaccuracy.

XX. STRAIN GAUGE ANALYSIS

Buckle strain gauges have been used to estimate the strain rate and stiffness in tendons of horses both in amputation specimens[64] and in the intact animal and compared with bone strain by Barnes and Pinder[10] (Figure 11). The principle is that a tendon weaved through the buckle, will on contraction cause it to deform, and this change will be measured by strategically placed gauges. Strain gauges have been applied to human knee ligaments by Kennedy et al.[78] Mercury-filled silicon rubber tubes were attached in parallel with the ligament under investigation; the resultant stretching of the rubber altered the electrical resistance of the mercury.

Both of these types of gauge are inaccurate in use. However, it is hoped that regenerated anterior cruciate ligaments will be investigated, using buckle strain gauges or some other method to assess the functional repair as well as the structural. No one has yet really solved the problem of directly measuring the strain in anterior cruciate ligaments, because it is difficult to apply any type of gauge accurately without it shifting on its ligamentous attachments under load. The deep-lying situation of the cruciate ligament is an added difficulty. Patients with knee instability are being examined using various jigs (Figure 12).[87] These devices make it possible to record degree of joint laxity. There are other methods. The patients who are considered for knee replacement in our unit undergo gait analysis. The method of measurement is as follows: the forces under the foot are accurately measured, using a force plate with transducers. The angle of the knee in the coronal and saggital planes

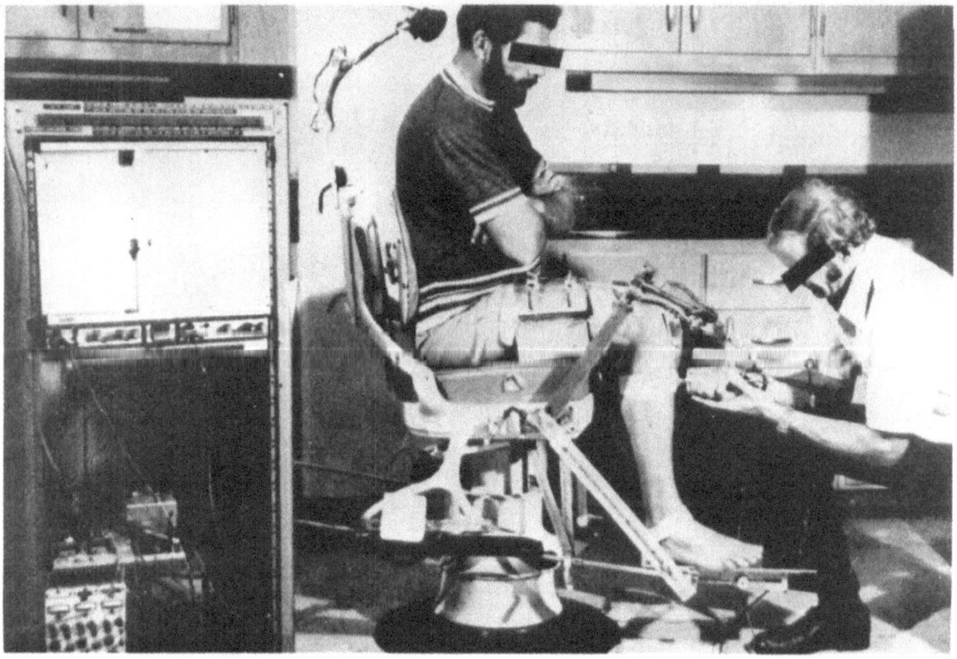

A

B

FIGURE 12. Jig used for testing knee joint stability. (A) In AP plane, (B) in lateral plane. (From Markoff, K. L., Graff-Radford, A., and Amstutz, H. L., *J. Bone J. Surg.*, 60A, 664, 1978. With permission.)

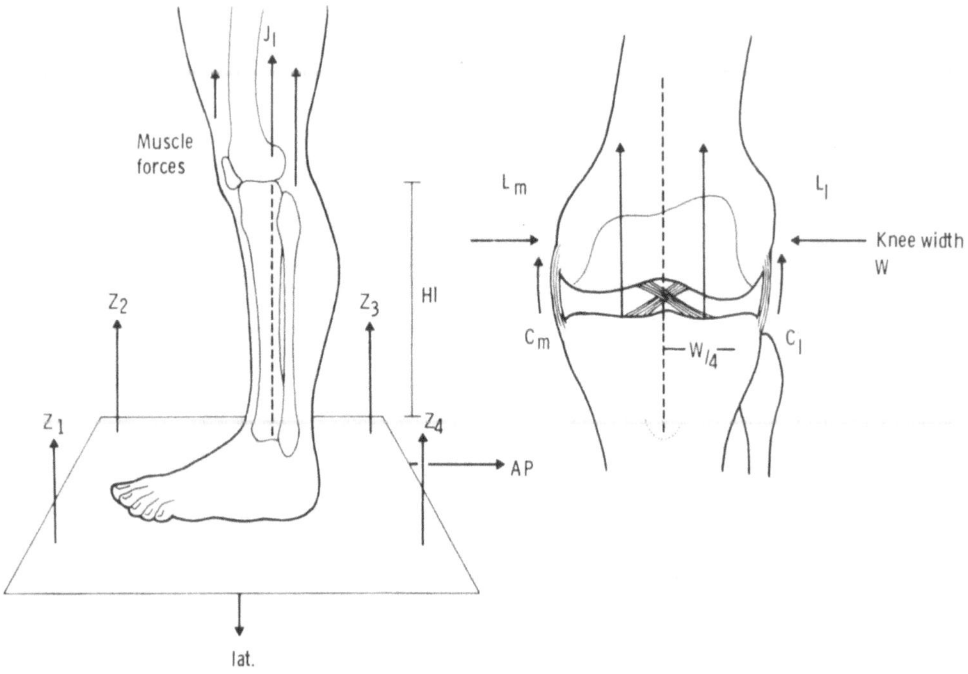

FIGURE 13. Method of using a force plate to analyze load at the knee joint. (From Johnson, F., Scarrow, P., and Waugh, W., *Med. Biol. Eng. Comput. J.*, Pergamon Press, 1981. With permission.)

is measured when the foot is placed on the force plate. It is then possible to calculate the relative amount of the weight passing through the medial plateau of the tibia and the load taken by the collateral ligaments. The total joint force is also calculated[72] (Figure 13). We think that the percentage of medial plateau load thus measured is reasonably accurate, though estimation of the total joint force and collateral ligament load during the stance phase of walking is probably not accurate enough to be useful clinically.

The ultimate strength of the anterior cruciate ligament has been measured by longitudinal stress by several writers.[97] Recently, Butler et al.[26] altered this jig so that stress is taken in a horizontal and slightly more physiological plane with reference to the femur. An assessment of strength to rotational stress has not yet been reported. The fibers do undoubtedly spiral, and one of the functions of the anterior cruciate is to resist anterolateral rotation.[77]

XXI. SUMMARY OF LIGAMENT STRESS

This discussion of the anterior cruciate ligament shows that the most important structural aspect of any ligament relates to its gross anatomy rather than to its fine structure. Of course, studying the fiber which supplies the strength can lead to some understanding of how it functions. It is such research that may elucidate the abnormalities in collagen which may predispose to failure.

Hence, in studying ligaments macroscopically, it is necessary to understand the gross anatomy, complementing it with dissections and microscopical examination of fiber orientation. Having done this, one can have some idea of its function and can plan the relevant tests required around the joint. Most ligaments will be less complex than the anterior cruciate!

XXII. RECONSTRUCTION OF ANTERIOR CRUCIATE

The anterior cruciate ligament when ruptured is very difficult to repair and frequently fails. In athletes there may be no symptoms of rupture.[30,92] As other ligaments are often injured at the same time as the anterior cruciate, complex instability syndromes are seen clinically.[42,80] Although some deny that isolated division of the anterior cruciate ligament ever leads to clinical instability, the general feeling is now that it does.[52,77] The variety of operations to overcome the instability by both intracapular and the extracapular reconstructions[36,69,73,128] suggest that no reconstruction accurately recreates the effect of the normal ligament — even those which transfer part of the patella tendon across the joint near to the normal cruciate position. The surgical problem is that accurate length and position of the reconstruction is difficult to achieve.

Recreation of the normal gross anatomy is easy to understand. It has been argued that extra-articular repairs are more effective; this is because the attachments of the restraint to abnormal movement are as far as possible away from the center of the joint, hence the increasing effective moment.[36] Some, including Thomson et al.,[128] have advised that a dynamic sling, using muscle transfers, produces a stabilization which is adjustable and has its maximal effect when the knee is straight during walking.

At least one total knee replacement requires the excision of both cruciates and is without any built-in constraint, yet is perfectly stable.[47] This is achieved by placing the components in the knee as tightly as possible and by accurately placing the tibial component so that when the patient walks this component is parallel to the ground.

XXIII. ANATOMY OF FLEXOR TENDONS

The ultimate strength of a tendon is not really as important as it is for a ligament. Nevertheless, it has been measured.[66] More important is the gliding function of ruptured tendons within the synovial sheaths.[134] Movement was restricted in a repaired tendon after one week. As an example, in the hand each digit has two flexor tendons, each arising from its appropriate muscle in the distal 1/3 of the forearm and inserted into the middle phalanx, in the case of the sublimus, and the distal phalanx, in the case of the profundus. From the proximal palmar crease (opposite the metacarpal neck) to the base of the middle phalanx, these tendons move within a synovial sheath consisting of an inner gliding synovial layer and an outer retaining fibrous layer. This is thickened over the middle of each phalanx to form pulleys to prevent bowstringing of the tendons, and therefore increasing the efficiency of grip. The synovial layer secretes fluid similar to that in synovial joints, which is principally hyaluronic acid in saline. The blood supply to these tendons[85,143] is derived from several vessels arising from a plexus of vessels around the sheaths. These vessels cross the synovial space and enter the sublimus tendon and from there, the profundus tendon. These are the vincular vessels (Figure 3).

XXIV. TENDON SHEATH REPAIR AND DEFECTS OF TENDON GLIDING

Greenlee[55] noted that sheath repair was delayed with respect to tendon repair by about one week, the tendon cells being originally unlined. Within sheath development he noted also the re-establishment of the vincular circulation.

Because of the tight inextensible sheath, any thickening of either tendon or synovium may cause obstruction to the free running of the tendon and cause a "triggering effect" when the tendon sticks at one point of its excursion and may be unable to move further.

Obstruction is common after injury, and this is noticed principally after devascularization of the tendon.[91] When an avascular tendon graft is taken from elsewhere in the body and inserted into the bed of the original tendon, it results in a length of dead tendon. Naturally this tendon must regain its blood supply; this is attained through adhesions which pass from the sheath into the tendon, causing it to stick down.[107] It has been shown that isolated segments of tendon receive nourishment from synovial fluid without any blood supply.[87] Matthews and Richards[90] performed several experiments with different degrees of tendon damage and showed that adhesions occurred after summation of several smaller injuries, particularly devascularizing sutures combined with immobilization. Attempts to reduce adhesions[121] have shown some benefit. However, the tendon graft placed in the pseudosynovial sheaths produced by a silastic rod has to obtain its blood supply from somewhere. Therefore, after tendon repair one occasionally needs to release a tendon from its sheath by a surgical operation. This works satisfactorily provided the original blood supply is adequate, for release of a tendon graft is to invite rupture through structural weakening after a second episode of devascularization. Methods to reduce adhesions include the injection of steroids around a tendon to prevent inflammatory reactions. In view of Potenza's work we must be sure that we do no harm. One possible way to improve the results of tendon repair would be to improve the bed of the tendon rather than the tendon itself. Eskeland et al.[37] showed that after a silicon rubber implant a pseudosheath was formed indistinguishable from the original; and this was also reported by Takasugi.[126] However, this operation requires sacrifice of the tendon and later, after the sheath has reformed, grafting of a tendon from elsewhere in the body. Attempts to use various biological and synthetic materials as a wrap around the original tendon have not been entirely successful.[121]

XXV. ALTERATIONS IN TENDONS AND LIGAMENTS IN SPECIAL SITUATIONS

A. Injury

When dense connective tissue is injured, the response is similar to the development of an embryonic tendon. If repair occurs with the ruptured ends of a tendon or a ligament apart, then ultimate function may be poor. Opposition of the two ends should therefore be achieved either by direct suture or by positioning within a splint.

Rupture is most likely to occur at the narrowest part of the tendon. In this connection Zielinki,[146] by measuring the tensile strength of the posterior cruciate ligament in dogs, showed that its strength was related to the thickness of the ligament.

B. Repair

After injury the first phase of repair is one of removing dead tissue and blood clot from the defect by an acute inflammatory reaction followed by early granulation tissue from the surrounding connective tissue.[6] This contains many new blood vessels. The cellular response with fibroblasts and gradually increasing collagen fiber function requires function of the repaired tissue for its maturation to that of a relatively acellular ligament. A similar response is seen at the ends of a freeze-dried tendon graft;[107] however, if the graft itself remains intact no evidence of creeping substitution is seen. A similar reaction has been achieved with autologous tendon grafts.

C. Artificial Materials as Prosthesis

Even where a gap is closed by artificial materials, a similar reaction takes place provided the material is sufficiently inert.[70,141] However, functional stress is essential for the development of the induced tissue into ligamentous material[41] with approximately the same strength as the original.[40] Walker[141] examined regenerated tendons biomechanically, using the Instron® machine. He confirmed that using polyester mesh, as with carbon fiber, the replacement ligament was much larger than the original. In our study it was approximately six times the normal diameter.[71] The tendons were 15% less stiff than normal and their elastic moduli were lower. The volume and speed of regeneration also depends on the bed in which the implant lay; and the more connective tissue available for regrowth the better.

XXVI. IMPORTANCE OF FUNCTION IN REPAIR AND REGENERATION

Simple excision of the muscle in a young animal caused the tendon to thin to 85% of its previous size.[116] Tendon development in the young animal did not proceed when the joints across which it acted were fixed. Neither did they grow when they were defunctioned. Scothorne also confirmed our findings that during repair a gap in a tendon was filled in by tissue indistinguishable from the original, provided tension was present. Repair was prevented by the absence of stress.

As a result of these findings we have departed from the period of immobilization traditionally advocated to management with a more rapid course of mobilization using some form of moveable splint.

XXVII. ALTERATIONS IN COLLAGEN FIBER AND TYPE

Stevens et al.[124] showed that denaturation of collagen fibers occurred during mechanical rupture. In scanning electron micrograph studies, fibers were seen to taper markedly, leading up to a coiled knot at the point of rupture which is typical of denaturation and is digestible by trypsin. They recommended digestion prior to repair!

Alteration in collagen type has been recorded in patients with Achilles tendon ruptures which are explored soon after injury. Coombs et al.[33] showed that some tendons contained excess Type III collagen. They related this to the likelihood of injury, explaining that it was less able to resist tensile forces than Type I because of alterations in crosslinking. Though this may be the case, it has also been suggested that these patients sustain microruptures within the Achilles tendon before complete rupture. This is supported by finding healing areas within the tendon soon after complete rupture, before this in itself could lead to any response.

Elongation of a normal Achilles tendon leads to sarcomere adaptation[127] with a resultant muscle shortening decreasing the effect of the tendon lengthening on the whole muscle tendon unit. Postachini et al.[106] suggest that contraction of tendons during repair may be due to the presence of actin contractile elements in regenerating tendons. This aspect is insufficiently studied as yet. The presence of myofibroblasts in tissue excised from patients with progressive Dupuytren's contracture seems to confirm the importance of this observation. This lesion has not been shown to be caused by a single factor, although it has been suggested that repeated injury from constant vibration such as a pneumatic drill may contribute. Type III collagen was also shown in greater quantity in Dupuytren's tissue.[8]

A. Force Inducing Injury

The force required to rupture human patellar tendons is known. During a weight-lifting competition filmed for training purposes, one competitor suffered this injury which occurred at 550 to 560 nm with a knee at an angle of 89.2° while in the jerk phase of lifting.[145]

B. Effects of Aging

There are numerous publications relating to the aging of collagen and of ligaments them selves. Viidik[136] showed that in rat tail tendon, thermal contraction-relaxation and dissolution altered with age, the contraction phase being much more evident.

C. Chemical Changes

These changes have been related to alterations in crosslinking.[138] The bone ligament junction has also been shown to alter with age by Tipton et al.[129] He found an increasing amount of hydroxyproline with age, and that breakdown occurred in 70% at the bone interface and 30% in the non-mineralized cartilage of the ligament itself. Because at their insertion, collagen fibers are intimately associated with glycosaminoglycan, these changes cannot be due to collagen alone. He considered that the alteration lay in the polymerization of ground substance and links between GAG and collagen. However Bentley[13] pointed out some of the difficulties in measuring the GAG fraction because it is difficult to separate from collagen. Heating the collagen he thought chemically reduced the hydrogen bonds. He concluded that thermal stabilization occurred with age, increasing the number of hydrogen bonds (which are the only ones measureable using borohydride) in the intermolecular crosslinks which involve hydroxylation and deamination of lysine groups.

Miyahara et al.[93] did not find any changes in hydroxylation with age. He did however, find that aldehyde crosslinks were more common in the older age groups. These experiments used skin but, as stated before, the collagen fibers of skin and tendon are similar.

XXVIII. MECHANICAL STRENGTH

Vogel[140] showed increasing strength of connective tissue of various organs, including rat skin and tail tendon, until maturation; a decreased strength occurred during senescence. During growth there was a rise in ultimate strain and load, tensile strength, and modulus of elasticity paralleled by collagen changes, but the GAG content reduced gradually over the whole period particularly during maturation.

Nathan et al.[96] tested isolated tendon fibers (from rat tail) on a specially prepared Instron® machine. He showed increasing strength to 6 months of age but stability thereafter. A change of fibrils in Type I collagen from unimodal to bimodal by maturity was mirrored by a reversion to the unimodal state after tendon rupture had occurred.[100]

XXIX. FIBRIL SIZE

Similarly fibril size was shown to have an effect on strength and also alter with age.[100] Large collagen fibers should have greater strength because of the increased number of intrafibrillar crosslinks. To withstand creep, a higher surface/unit area with increased interfibrillar crosslinks is needed. He concluded that an ideal collagen fiber diameter must exist. He suggested that fibrils had an average diameter of between 1500 to 5000 Å and ultimate strength of 60 MPa. During stress, the metabolism of cells controlling the glycosaminoglycans was stimulated; as these were determined by the ultimate collagen fiber size, this was also influenced.

XXX. ELASTIC FIBERS

Similar experiments on elastic fibers[118] showed that the elastin/collagen ratio is greater at 10 weeks of age in females (0.055:0.032). Therefore the risk of congenital dislocation of the hips is increased. Elastin was also noted to be increased in women, especially following estrogen therapy.

Age increased the density and aggregation of elastin fibrils. The mature elastic fibers were closely related to collagen bundles although not necessarily in the same direction.

XXXI. DISEASES

Alteration of collagen is seen regularly with many diseases. Perhaps the most obvious is Ehler's Danlos Syndrome Type V. Because of a reduction in lysyl oxidase enzyme,[120] crosslinking is normal; but the links formed lack the tight packing of normal collagen fibrils, a feature confirmed by electron microscope. Others have suggested similar enzymic defects: Type VI with reduced lysyl hydroxylase, and Type VII with reduced procollagen peptides.

XXXII. OTHER FORMS OF LIGAMENT LAXITY

Ligamentous laxity was seen to be paralleled by scar spreading after surgery for correction of recurrent dislocation of the joint,[86] particularly when generalized laxity is the underlying cause. Congenital dislocation of the hip was associated with a more generalized ligamentous laxity, both in the patients and their parents,[28,144] and the extractable collagen and collagen content was lower in these patients than normal.[45]

Other generalized diseases, such as muscular dystrophy,[125] are associated with poor collagen alignment. As this was an early factor Stinson[125] thought it to be a primary rather than secondary defect. In idiopathic club foot, collagen synthesis is increased above normal, and this is also a feature of arthrogryphosis.[67] These findings, however, are investigated principally in the associated muscles rather than the tendons themselves.

Assessment of ligament laxity has been described recently by Burwell et al.[22] They used a new trigonometric method involving calculation of the angle to which the thumb could be pressed manually towards the radial border of each forearm. This is a modification of the methods previously described.[28]

XXXIII. SCOLIOSIS

Idiopathic scoliosis comprises another group of patients who have been shown to have collagen changes. This common condition consists of curvature of the spine, for as yet no obvious reason. As there are definite epidemiological groups in which disease progresses specifically at the growth points of development, it is not surprising that they have a number of other defects including ligamentous laxity.[22,25]

Intervertebral disks removed at operation[25] showed alteration in collagen type but not in the amount throughout the spine in scoliosis. They concluded that reducible crosslinks are greater in these patients than in normal people. These findings have also been observed in chickens with scoliosis.[109]

Bushell[23] suggested that collagen extracts from disks showed a predominance of Type I over Type II and that this affects the collagen terminals reducing stability. Extraction of polymeric collagen from the skin of patients with scoliosis, showed less stability than control specimens, and this was associated with rapid growth. The finding of generalized collagen defects confirms the clinical impression of a multiple defect disease. Similar examination of skin in these patients shows a temporary decrease in collagen stability which became normal in adult life.[43]

After experimental posterior spinal fusion without actually touching the disk, no change occurred in the collagen of animals but did in humans, suggesting that this effect is secondary and stress related.[24] Bushell's finding of alteration in the amount of collagen at the apex of the curve only, and altered collagen type throughout the spine suggests that the effect is primary.

XXXIV. EFFECTS OF ACTIVITY

The other major factor altering collagen is the effect of activity and immobilization. Vallas et al.[132] trained rats which were either normal or hypophysectomized, to increase fitness. However, his measurements of activity depended on the oxygen utilization of the tendon, which is rather low. Although changes were seen in muscle with the induced increased fitness, no tendon alterations were observed.

Akeson et al.[1] showed that limb immobilization caused loss of proteoglycan and up to 10% of normal collagen. However, increased torque was required to extend an immobilized knee, supporting a theory of more than a simple loss of lubrication. There was a significant increase in reducible inter-molecular crosslinks.

Claus-Walker et al.[31] similarly failed to show any increase in the urinary excretion of collagen degradation products during weightlessness although quadriplegic patients did lose collagen.

Denervation of limbs was accompanied by loss of collagen from many types of connective tissue, especially ligaments.[79] Ligaments, however, also had a greater replacement of collagen. Unfortunately the new collagen was not typed.

XXXV. INJURY IN ABNORMAL TISSUE

Rupture following minor injury is a common result of unaccustomed exercise. The common sites are the supraspinatus tendon at the shoulder, the Achilles tendon, and the quadriceps tendon above the patella. The tendons involved are thought to be degenerate.[5] Degenerate tendon occasionally becomes calcified, and this may be noted on x-ray.

XXXVI. SPONTANEOUS RUPTURE

Finally, spontaneous tendon ruptures are associated with several generalized diseases. That most commonly quoted occurs in patients with chronic renal failure.[95] All patients had 2° hyperparathyroidism and had been on long term hemodialysis. The hyperparathyroidism appears to affect the tendosseous junction.

XXXVII. EFFECTS OF DRUGS AND VITAMINS

One of the effects of scurvy (lack of vitamin C) is poor wound healing or even disruption of scars already healed.[32] However, the underlying defect is more obscure. Although many claim to measure collagen synthesis by prolylhydroxylase activity, Chvapil states that this is not an accurate measurement, being independent of collagen volume. The research on Vitamin C is on skin with little reference to tendons, since these clinically are rarely affected. The actual vitamin effect on dense connective tissue is not known since, although Kao et al.[75] showed a rise in synthesis after the addition of Vitamin C, they found this to be a direct effect rather than through the increased activation of prolylhydroxylase.

The drugs most commonly given to affect collagen are glucocorticosteroids. In organ culture, cortisol increased synthesis when given for the last 3, 24, or 48 hr of a 96 hr experiment.[35] When given for the whole time, synthesis decreased and the effects could be blocked by the cortisone antagonist cortexolone. This was said to suggest that physiological cortisone may have an all-or-none effect, maintaining differentiation and collagen synthesis but inhibiting the growth of precursor cells.

Direct injection into a tendon, contrary to clinical impressions,[27] did not cause tendon rupture in normal tendon; Matthews et al.[89] suggested that rupture is probably due to some underlying disease. Besides discussing the effects of aging on collagen, Viidik[138] mentions that lysyl oxidase, and thus crosslinking, is affected by β-amino-proprionitrile, resulting in a lathyrism. This is a disease characterized by joint dislocation and aneurisms due to collagen degeneration;[17] thus suggesting that this drug can be useful in the short-term treatment of constrictive lesions such as stricture, while they are healing but not when established.

Different interference with the crosslinking of collagen allows D-penicillamine to be useful in rheumatoid arthritis and sclerodema.

Thyroid hormones and the pituitary-adrencortical system have profound affects on collagen. Thyroidectomy or suppression of thyroid activity reduces the amount of collagen formed; however, the insoluble collagen fraction becomes greater than normal so that ultimate strength and elastic stiffness are increased. Corticosteroids were noticed similarly to increase these factors whether given directly or stimulated indirectly by ACTH.

XXXVIII. CONCLUSIONS

1. Tendons and ligaments are types of dense connective tissue. Their gross anatomical features are more important from a functional standpoint than is their fine structure as revealed by electron micrography.
2. The change in collagen with disease needs to be investigated further in order to prevent the effects of disease, possibly by substitution of missing enzymes.
3. Reconstruction of injured joints is in its infancy, as is shown by the multitude of options open to a surgeon to deal with any given condition and by the lack of any general agreement on the effects of injury.
4. The regeneration of ligaments across a gap bridged by an artificial material such as carbon fiber is promising and is now being evaluated in patients. Its ultimate place has yet to be decided. However, its potential promise will not be fully known for several years. This is because after operation, neither repair nor reconstruction is static, and the gradual lengthening of the repair tissue remains as a disappointment to both patient and surgeon.

However, it behooves us to understand more of the normal structure of the material we are replacing, and the mechanical properties and biological effect of what we are replacing it with. In this way, our operative surgery may be of the highest available standard and be more effective. Only in this scientific way can we approach the clinical problem with any hope of lasting success.

ACKNOWLEDGMENTS

I would like to thank Professor R. G. Burwell for his help and encouragement in the preparation of this manuscript, and Miss S. Tippett for her unfailing energy in typing it.

REFERENCES

1. **Akeson, W. H., Armiel, D., Mechanic, G. L., Woo, S. L-Y., Harwood, F. L., and Hamer, H. L.,** Collagen crosslinking alterations in joint contractures; changes in the reducible crosslinks in perianticular connective tissue collagen after nine weeks immobilisation. *Connect. Tissue Res.*, 5, 15, 1977.
2. **Allain, J. C., Lebus, M., Bazin, S., Bailey, A. J., and Delaunay, A.,** Isometric tension developed during heating of collagenous tissue: relationships with collagen cross-linking. *Biochim. Biophys. Acta*, 563(1), 147, 1978.
3. **Anderson, J. C. and Labedz, R. I.,** Extractability of connective tissue glycoprotein from bovine tendon. *Biochem. Soc. Trans.*, 5(2), 434, 1977.
4. **Anderson, J. C., Labedz, R. I., and Kewley, M. A.,** Effect of bovine tendon glycoprotein on the formation of fibrils from collagen solutions. *Biochem. J.*, 167, 345, 1977.
5. **Apley, A. G.,** *A System of Orthopaedic and Fractures*, 4th ed., Butterworths, Reading, Mass., 1973.
6. **Arnoczky, S. P., Rubin, R. M., and Marshall, J. V. L.,** Microvasculature of the cruciate ligament and its response to injury. *J. Bone J. Surg.* 61A, 1221, 1979.
7. **Bachinger, H. P., Bruckner, P., Timpl, R., and Engel, J.,** The role of cis- trans isomerization of peptide bonds in the coil — triple helix conversion of collagen. *Eur. J. Biochem.*, 90(3), 605, 1978.
8. **Bailey, A. J., Sims, T. J., Gabbiani, C., Bazin, S., and Lelous, M.,** Collagen of Dupuytrens Disease, *Clin. Sci. Mol. Med.* 53, 499, 1977.
9. **Barenberg, S. A., Filisko, F. E., and Geil, P. H.,** Ultra structure deformation of collagen. *Connect. Tissue Res.* 6, 25, 1978.
10. **Barnes, G. R. G. and Pinder, D. N.,** In vivo tendon tension and bone strain measurement and correlation. *J. Biomech.*, 7, 35, 1974.

11. **Bell, E., Ivarsson, B., and Merrill, C.,** Production of tissue-like structure by contraction of collagen lattices by human fibroblasts of different proliferation potential in vitro. *Proc. Nat. Acad. Sci.*, 76(3), 1274, 1979.
12. **Belton, J. C., Micheali, D., and Fudenberg, H. H.,** Freeze etch study of collagen, *Arthritis Rheum.*, 18, 443, 1975.
13. **Bentley, J. P.,** Ageing of Collagen, *J. Invest. Dermatol.*, 73, 80, 1979.
14. **Bichat, M. F. X.,** *General Anatomy Applied to Physiology and the Practice of Medicine*. 1802 Translated by Coffryn, C., Rev. Colvert, G. in 1824., Shackell, Arrowsmith.
15. **Binder, M. D., Kroin, J. S., Moore, G. P., and Stuart, D. G.,** Response to Golgi tendon organs to single motor unit contractions. *J. Physiol.*, 271, 337, 1977.
16. **Bloom, I. and Fawcett, D. W.** *A Textbook of Histology*, 9th ed., W. B. Saunders, Philadelphia, 1968.
17. **Bondy, P. K. and Rosenberg, L. E.,** *Metabolic Control and Disease*, W. B. Saunders, Philadelphia, 1980, 111.
18. **Boueck, R. J., Noble, N. L., and Gunga-Smith, Z.,** A possible role of dihydrodihydroxylysin — onorleucine in collagen fibre and bundle formation. *Biochem. J.*, 177, 853, 1979.
19. **Bowitz, R., Jonak, R., Nemetschek-Gansler, H., Nemetschek, Th., and Riedl, H.,** The elasticity of the collagen triple helix. *Naturwissen schaften* 63(12), 580, 1976.
20. **Brantigan, O. C. and Voshell, A. F.,** The mechanics of the ligaments and menisci of the knee joint. *J. Bone J. Surg.*, 23A, 44, 1941.
21. **Bruckner, P., Bachinger, H. P., Timpl, R., and Engel, J.,** Three conformationally distinct domains of amino-terminal segment of type III procollagen and its rapid triple helix coil formation. *Eur. J. Biochem.*, 90, 595, 1978.
22. **Burwell, R. G., Dangerfield, P. H., and Vernon, C. L.,** Bone asymmetry in joint laxity in the upper limbs of children with adolescent idiopathic scoliosis. British Association of Clinic Anatomists, Summer, 1980. *Ann. Royal College Surgeons of England*, 63, 209, 1981.
23. **Bushell, G. R., Ghosh, P., and Taylor, T. K. E.,** Collagen defect in idiopathic scoliosis *Lancet*, 2(8080), 94, 1978.
24. **Bushell, G. R., Ghosh, P., Taylor, T. K. F., Sutherland, J. M., and Braund, K. G.,** The effect of spinal fusion on the collagen and proteoglycans of the canine intervertebral disc. *J. Surg. Res.*, 25, 61, 1978.
25. **Bushell, G. R., Ghosh, P., Taylor, T. K. F., and Sutherland, J. M.,** The collagen of the intevertebral disc in adolescent idiopathic scoliosis. *J. Bone J. Surg.*, 61B, 501, 508, 1979.
26. **Butler, D. L., Noyes, F. R., and Grood, E. S.,** Ligamentous restraints to anterior-posterior drawer in the human knee. *J. Bone J. Surg.*, 62A, 259, 1980.
27. **Cargill, A. O'R.,** Closed rupture of the Achilles tendon. *Brit. J. Hosp. Med.*, 524, 1976.
28. **Carter, C. and Wilkinson, J.,** Persistent joint laxity and CDH. *J. Bone J. Surg.*, 468, 40, 1964.
29. **Chatterjee, T. K., Spadaro, J. A., and Vook, R. W.,** Surface model of the collagen fibril as determined by two stage plastic replica. *Experienta* 34 (12), 1635, 1978.
30. **Chick, R. C. and Jackson, D. W.,** Loss of the anterior cruciate ligament in young athletes. *J. Bone J. Surg.*, 60A, 970, 1978.
31. **Claus-Walker, J., Singh, J., Leach, C. S., Hatton, D. V., Hubert, C. W., and DiFerrante, N.,** Urinary excretion of collagen degradation products by quadriplegic patients and during weightlessness. *J. Bone J. Surg.*, 59A, 209, 1977.
32. **Kohen, I. K. and Keiser, H. R.,** Disruption of healed scars in scurvy: the result of disequilibrium in collagen metabolism. *Plast. Reconst. Surg.*, 57, 376, 1976.
33. **Coombs, R. R. H., Klenerman, L., Narcisi, P., Nichols, A., and Pope, F. M.,** Collagen typing in Achilles tendon rupture. *J. Bone J. Surg.*, 62B, 258, 1980.
34. **Davis, N. R., Risen, O. M., and Pringle, G. A.,** Stable non-reducible crosslinks of mature collagen. *Biochemistry*, 14(9), 2031, 1975.
35. **Dietrich, J. W., Canalis, E. M., Maina, D. M., and Raisz, L. G.,** Effects of glucocorticoids on foetal rat bone collagen synthesis in vitro. *Endocrinology* 104, 715, 1979.
36. **Ellison, A. E.,** Distal iliotibial band transfer for anterior lateral rotatory instability of the knee. *J. Bone J. Surg.*, 61A, 330, 1979.
37. **Eskeland, G., Eskeland, T., Hevis, T., and Teigland, J.,** The ultrastructure of the normal digital flexor tendon sheath and of the tissue formed around silicone and polyethylene implants in man, *J. Bone J. Surg.*, 59B, 206, 1977.
38. **Falcon, J.,** (1649) in Gratz, C. M., History of Tendon Suture, *Med. J. Rec.*, 127, 156, 1928.
39. **Flint, M. H.,** Basis of the histological demonstration of tension in collagen. The ultrastructure of collagen, in *International Symposium and Workshop Relation of the Ultrastructure of Collagen to the Healing of Wounds*. Longacre, J. J., Ed., Charles C Thomas, Springfield, Ill., 1969.
40. **Forster, I. W.,** A study of the mechanism by which carbon fibre acts as a tendon prosthesis. *J. Bone J. Surg.*, 68B, 376, 1976.

41. **Forster, I. W., Ralis, Z. A., and Jenkins, D. H. R.,** Biological reaction to carbon fibre implants: the formation and structure of carbon fibre induced "Neotendon". *Clin. Orthop. Relat. Res.*, 131, 299, 1978.
42. **Fowler, P. J.,** The classification and early diagnosis of knee joint instability. *Clin. Orthop. Relat. Res.*, 147, 15, 1980.
43. **Francis, M. J. O., Smith, R., and Sanderson, M. C.,** Collagen abnormalities in idiopathic adolescent scoliosis. *Calcif. Tissue Res.*, 22 Suppl. 381, 1977.
44. **Frazer, R. D. B. and MacRae, T. P.,** The crystalline structure of collagen fibrils in tendons. *J. Mol. Biol.*, 127, 129, 1979.
45. **Fredensborg, N. and Uden, A.,** Altered connective tissue in children with congenital dislocation of the hip. *Arch. Dis. Child.*, 51, 887, 1976.
46. **Frederickson, R. G., Morse, D. E., and Low, F. N.,** High voltage E M of extracellular fibrillogenesis. *Am. J. Anat.*, 150(1), 1, 1977.
47. **Freeman, M. A. R., Insall, J. N., Besser, W., Walker, P. S., and Hallel, T.,** Excision of the cruciate ligaments in total knee replacement. *Clin. Orthop. Relat. Res.*, 126, 209, 1977.
48. **Fujii, K., Tanzer, M. L., and Cooke, P. H.,** Collagen fibrogenesis and formation of complex crosslinks. *J. Mol. Biol.*, 106, 223, 1976.
49. **Gabbiani, G., Hirschel, B. J., Ryan, G. B., Statkov, P. R., and Majno, G.,** Granulation tissue as a contractile organ. *J. Exp. Med.*, 135, 719, 1972.
50. **Gelberman, R. H., Amiel, D., Rudolph, R. M., and Vance, R. M.,** Dupuytren's Contracture: an electron microscopic biochemic and clinical correlative study. *J. Bone J. Surg.*, 62A, 1980.
51. **Gelman, R. A., Williams, B. R., and Pieg, K. A.,** Collagen fibril formation: evidence for a multistep process. *J. Biol. Chem.*, 254(1), 180, 1979.
52. **Girgis, F., Marshall, J. L., and Al Monajem, A. R. S.,** The cruciate ligaments of the knee joint. *Clin. Orthop. Rel. Res.*, 106, 216, 1975.
53. **Goldberg, B.,** Binding of soluble Type I collagen molecules to fibroblasts in plasma membranes. *Cell,* 16(2), 265, 1979.
54. **Warwick, R. and Williams, P. L.,** Eds., *Gray's Anatomy,* 35th ed., Longman Group, Saunders, Philadelphia, 1973.
55. **Greenlee, T. K., Beckham, L., and Pike, D.,** A fine structural study of the development of chick flexor digital tendon: a model for synovial sheathed tendon healing. *Am. J. Anat.*, 143, 303, 1975.
56. **Gregory, J. E., Harvey, R. J., and Proske, U.,** A "late supernormal period" in the recovery of excitability following an action potential in muscle spindle and tendon organ receptors, *J. Physiol.*, 271, 449, 1977.
57. **Groves, E. W.,** The cruciate ligaments of the knee joint. Their function, rupture and the operative treatment of the same. *Brit. J. Surg.*, 7, 505, 1920.
58. **Guillemean, J.,** (1598) in Gratz, C. M., History of Tendon Structure, *Med. J. Rec.*, 127, 156, 1928.
59. **Hahn, L-H. E. and Yamada, K. M.,** Identification of a collagen binding fragment of the adhesive glycoprotein fibronectin. *Proc. Natl. Acad. Sci. U.S.A.*, 76(3), 1160, 1979.
60. **Haines, R. W.,** A note on the actions of the cruciate ligaments of the knee joint. *J. Anat.*, 75, 373, 1941.
61. **Von Haller, A.,** *Von den Emgfindlichendlichen und Reizbaren Teilen des Menschlichen.* Korpess, Leipzig, 1752.
62. **Ham, A. W. and Cormack, D. H.,** *Ham's Histology,* 8th ed., J. B. Lipincott, New York, 1979.
63. **Heathcote, J. G. and Al-Alawi, S.,** Osteogenesis imperfecta: evidence for the existence of an abnormal amino acid sequence in the molecule of dermal collagen. *J. Inherited Metabol. Dis.*, 1, 123, 1978.
64. **Herrick, W. C., Kingsbury, H. B., and Lou, D. Y. S.,** A study of the normal range of strain, strain rate and stiffness of tendon. *J. Biomed. Mater. Res.*, 12, 877, 1978.
65. **Hickey, D. S. and Hukins, D. W. L.,** Effect of methods of preservation on the arrangement of collagen fibrils in connective tissue matrices: an X-ray diffraction study of the annulus fibrosis. *Connect. Tissue Res.*, 6, 223, 1979.
66. **Holden, C. E. A. and Northmore-Ball, M. D.,** The strength of the profundus tendon insertion, *Hand* 7, 238, 1975.
67. **Ionesescu, V., Maynard, J. A., Ponsetti, I. V., and Zellweger, H.,** The role of collagen in the pathogenesis of idiopathic club foot: biochemical and electron microscopical correlations. *Helvetia Paediatrica Acta,* 29, 305, 1974.
68. **Ippolito, E., Natali, P. G., Postacchini, F., Accinni, L., and DeMartino, C.,** Ultrastructural and Immunochemical Evidence of Actin in Tendon cells, *Clin. Orthop. Relat. Res.*, 126, 282, 1977.
69. **Ireland, J. and Trickey, E. L.,** MacIntosh tendesis for anteriolateral instability of the leg. *J. Bone J. Surg.*, 62B, 340, 1980.
70. **Jenkins, D. H. R., Forster, I. W., McKibbin, B., and Ralis, Z. A.,** Induction of tendon and ligament formation by carbon implants. *J. Bone J. Surg.*, 59B, 53, 1977.
71. **Jenkins, D. H. R.,** Repairs of cruciate ligaments with flexible carbon fibre. *J. Bone J. Surg.*, 60B, 520, 1978.

72. **Johnson, F. J., Scarrow, P., and Waugh, W.,** Assessment of loads in the knee joint. *Med. Biol. Eng.*, (in press).
73. **Jones, K. G.,** Results of use of the central one-third of the patellar ligament to compensate for anterior cruciate instability. *Clin. Orthop. Relat. Res.*, 147, 29, 1980.
74. **Kao, W. W-Y., Flaks, J. G., and Prockop, D. J.,** Primary and secondary effects of ascorbate on procollagen synthesis and protein synthesis by primary cultures of tendon fibroblasts. *Arch. Biochem. Biophys.*, 173, 638, 1976.
75. **Kao, W. W-Y., Berg, R. A., and Prockop, D. J.,** Kinetics for the secretion of procollagen by freshly isolated tendon cells. *J. Biol. Chem.*, 252(23), 8391, 1977.
76. **Kastelic, J., Galeski, A., and Bear, E.,** The multicomposite structure of tendon. *Connect. Tissue Res.*, 6, 11, 1978.
77. **Kennedy, J. C., Weinberg, H. W., and Wilson, A. S.,** The anatomy and function of the anterior cruciate ligament. *J. Bone J. Surg.*, 56A, 223, 1974.
78. **Kennedy, J. C., Hawkins, R. J., and Willis, R. B.,** Strain gauge analysis of knee ligaments. *Clin. Orthop. Relat. Res.*, 129, 225, 1977.
79. **Klein, L., Dawson, M. H., and Heiple, K. G.,** Turnover of collagen in the adult rat after denervation. *J. Bone J. Surg.*, 59A, 1065, 1977.
80. **Larson, R. L.,** Combined instability of the knee. *Clin. Orthop. Related Res.*, 147, 68, 1980.
81. **Last, R. J.,** Some anatomical details of the knee joint. *J. Bone J. Surg.*, 30B, 683, 1948.
82. **Lew, W. D. and Lewis, J. L.,** A technique for calculating in vivo ligament lengths with application to the knee joint. *J. Biomech.*, 11, 365, 1978.
83. **Lillie, J. H., MacCallum, D. K., Scaletta, L. J., and Occhino, J. C.,** Collagen structure: Evidence for a helical organization of the collagen fibril. *J. Ultrastructure Res.*, 58, 134, 1977.
84. **Linsenmeyer, T. F., Gibney, E., Toole, B. P., and Gross, J.,** Cellular adhesion to collagen. *Exp. Cell Res.*, 116, 470, 1978.
85. **Lundborg, G., Murhage, R., and Rydevik, B.,** The vascularization of human flexor tendons within the digital synovial sheath region. *J. Hand Surg.*, 2, 417, 1977.
86. **MacNab, I. and MacNab, J.,** Ligamentous laxity and scar formation, *Clin. Orthop. Relat. Res.*, 135, 154, 1978.
87. **Markolf, K. L., Graft-Radford, A., and Amstutz, H. C.,** In vivo knee stability. *J. Bone J. Surg.*, 60A, 664, 1978.
88. **Mason, H. L. and Sheardon, C. G.,** The process of Tendon Repair. *Arch. Surg.*, 25(4), 615, 1932.
89. **Matthews, C. S., Sonstegard, D. A., and Phelps, D. B.,** A biomechanical study of rabbit patellar tendon: effects of steroid injection. *J. Sports Med.*, 2, 349, 1975.
90. **Matthews, P. and Richards, H. J.,** Factors in the adherence of flexor tendon after repair. *J. Bone J. Surg.*, 58B, 230, 1976.
91. **Matthews, P.,** Biological factors in the management of flexor tendon injuries, in *Recent Advances in Orthopaedics*, 3rd ed., McKibbin, B., Ed., Churchill Livingstone, Edinburgh, 1979, Ch. 9, 175.
92. **McDaniel, W. J. and Dameron, T. B.,** Untreated ruptures of the anterior cruciate ligament *J. Bone J. Surg.*, 62A, 696, 1980.
93. **Miyahara, T., Shiozawa, S., and Murai, A.,** Effect of age on amino acid composition of human skin collagen. *J. Gerontol.*, 33, 498, 1978.
94. **Mohanaradhakrishnan, V., Muthiah, P. L., and Hadhanhi, A.,** A few factors contributing to the mechanical strength of the collagen fibres *Arzneimittelforschung*, f:726. 1975.
95. **Morein, G., Goldschmidt, Z., Parker, M., Seelenfreund, M., Rosenfeld, J. B., and Fried, A.,** Spontaneous tendon ruptures in patients treated by chronic haemodialysis. *Clin. Orthop. Relat. Res.*, 124, 209, 1977.
96. **Nathan, H., Golgetter, L., Kobliansky, E., Goldschmidt-Nathan, M., and Morein, G.,** Energy absorbing capacity of the rat tail tendon at various ages. *J. Anat.*, 127, 589, 1978.
97. **Noyes, F. R., Delucas, J. L., and Torvik, P. J.,** Biomechanics of anterior cruciate ligament failure; an analysis of strain rate sensitivity and mechanism of failure in primates. *J. Bone J. Surg.*, 56A, 236, 1974.
98. **Noyes, F. R. and Grood, E. S.,** Knee ligament function and failure. Clinical applications of biomechanics presented at the American Orthopaedic Society of Sports Medicine, Las Vegas, Nevada, 1977.
99. **Oikarinen, A., Antinen, H., and Kivirikko, K. I.,** Further studies on the effect of collagen triple helix formation on the hydroxylation of lysine and glycosylations of hydroxylysine in chick embryo tendon and cartilage cells. *Biomech. J.*, 166, 137, 1977.
100. **Parry, D. A. D. and Craig, A. S.,** Collagen fibrils and elastic fibres in rat tail tendon; an E M investigation. *Biopolymers*, 17, 843, 1978.
101. **Parry, D. A. D., Craig, A. S., and Barnes, G. R. G.,** Tendon and ligament from the horse: an ultrastructural study of collagen fibrils and elastic fibres as a function of age, *Proc. Royal Soc. London*, (Biological Sciences) 203, 293, 1978.

102. **Parry, D. A. D., Barnes, G. R. G., and Craig, A. S.**, A comparison of the size distribution of collagen fibrils in connective tissues as a function of age and a possible relation between fibril size distribution and mechanical properties. *Proc. Royal Soc. London*, (BiologicalSciences) 203, 305, 1978.
103. **Parry, D. A. D. and Craig, A. S.**, E M evidence of an 80 Å unit in collagen fibrils. *Nature*, 282 (5735), 213, 1979.
104. **Petit, J. L.**, *Dictionary of Practical Surgery*, 6th ed., Cooper, London, 1830, 1077.
105. **Petkov, R.**, Ultrastructure of collagen fibrils. II. Evidence of the special organization of the fibril. *Anatomischer Anzeiger*, 144(5), 485, 1978.
106. **Postachini, F., Natali, P. G., Accinni, L., Ippolito, E., and DiMarino, C.**, Contractile filaments in cells of regenerating tendon. *Experientia* 33(7), 957, 1977.
107. **Potenza, A. D. and Melone, C.**, Evaluation of freeze dried tendon grafts in the dog. *J. Hand Surg.*, 3, 157, 1978.
108. **Proske, U. and Gregory, J. E.**, Multiple sites of impulse initiation in tendon organs. *Exp. Neurol.*, 50, 515, 1976.
109. **Riggins, R. J., Abbott, U. K., Ashmore, C. R., Rucker, R. B., and McCarrey, J. R.** Scoliosis in chickens. *J. Bone J. Surg.*, 59A, 1020, 1977.
110. **Risteli, L., Risteli, J., Salo, L., and Kivirikko, K. I.** Intracellular enzymes of collagen biosynthesis in 3T6 fibroblasts and chick embryo tendon and cartilage cells. *Eur. J. Biochem.*, 97, 297, 1979.
111. **Robins, S. P. and Bailey, A. J.**, The chemistry of collagen crosslinks. *Biochem. J.* 163, 339, 1977.
112. **Rojkind, M.**, Chemistry and biosynthesis of collagen. *Bull. Rheum. Dis.*, 30, 106, 1979.
113. **Ross, R., Fialkon, P. J., and Altman, K.**, The morphogensis of elastic fibres. *Adv. Exp. Med. Biol.*, 79, 7, 1977.
114. **Rozovskaya, T. R. and Yankovskaya, N. F.**, Sensory innervation of finger flexor tendons and their synovio-aponeurotic sheaths. *Acta Chir. Plast.*, 17(3), 125, 1975.
115. **Schaeffer, J. P.**, *Morris' Human Anatomy*, 10th ed., Blakiston, Philadelphia, 1933.
116. **Scothorne, R. J.**, The borderland of ergonomics and anatomy. *Ergonomics*, 18, 479, 1975.
117. **Serafini-Fracassini, A., Field, J. M., Smith, A. J., and Stephens, W. G. S.**, The ultrastructure and mechanics of elastic ligaments. *Advances in Experimental Medicine and Biology* 79, 97, 1977.
118. **Skikata, J., Sanada, H., Yamamuro, T., and Takeda, T.**, Experimental studies of the elastic fibre of the capsular ligament: influence of ageing and sex hormones on the hip joint capsule of rats. *Connect. Tissue Res.*, 7, 21, 1979.
119. **Shoshan, S. and Gross, J.**, Biosynthesis and metabolism of collagen and its role in tissue repair processes. *Israel J. Med. Sci.*, 10(5), 537, 1974.
120. **Siegal, R. C., Black, C. M., and Bailey, A. J.**, Crosslinking of collagen in the X linked Ehlers-Danlos type V disease. *Biochem. Biophys. Res. Commun.*, 88, 1979.
121. **Stark, H. H., Boyes, J. H., Johnson, L., and Asworth, C. R.**, The use of paratenon, polyethylene film or silastic sheathing to prevent restricting adhesions to tendons in the hand. *J. Bone J. Surg.*, 59A, 908, 1977.
122. **Stauffer, E. K. and Stephens, J. A.**, Tendon organs of cat soleus: static sensitivity to active force. *Exp. Brain Res.*, 23, 279, 1975.
123. **Stephens, J. A., Reinking, R. M., and Stuart, D. G.**, Tendon organ of cat medial gastrocenimus: response to active and passive forces as a function of muscle length. *J. Neurophys.*, 38, 1217, 1975.
124. **Stevens, F. S., Minns, R. J., and Finlay, J. B.**, Evidence of the local denaturation of collagen fibrils during mechanical rupture of human tendons. *Injury*, 6, 317, 1975.
125. **Stinson, R. H.**, Structural deterioration of tendon collagen in genetic muscular dystrophy. *Biochim. Biophys. Acta*, 400, 255, 1975.
126. **Takasuki, H., Inoue, H., and Akakori, O.**, Scanning electron microscopy of repaired tendon and pseudo sheath. *Hand*, 8, 228, 1976.
127. **Tardieu, Tabary, J-C., Tabray, C., and Huet-de-la-Tour, E.** Comparison of sarcomere number of in young and adult animals. *J. Physiol.*, Paris, 73, 1045, 1977.
128. **Thomson, S. K., Calver, R., and Monk, C. J. E.** Anterior cruciate ligament repair to rotatory instability; the Lindemann dynamic muscle transfer procedure. *J. Bone J. Surg.*, 60A, 917, 1978.
129. **Tipton, C. M., Matthews, R. D., and Martin, R. K.**, Influence of age and sex on the strength of the bone-ligament junctions in the knee joints of rats. *J. Bone J. Surg.*, 60A, 230, 1978.
130. **Trelsted, R. L. and Hayashi, K.**, Tendon collagen fibrillogenesis: intracellular subassociation and cell surface changes associated with fibril growth. *Dev. Biol.*, 71, 228, 1979.
131. **Trent, P. S., Walker, P. S., and Wolf, B.**, Ligament length patterns, strength and rotational axes of the knee joint. *Clin. Orthop. Relat. Res.*, 117, 263, 1976.
132. **Vallas, A. C., Tipton, C. M., Laughlin, H. L., Tcheng, T. K., and Matther, R. D.**, Physical activity and hypophysectomy on the aerobic capacity of ligaments and tendons. *J. Appl. Phys.*, 44, 542, 1978.
133. **Veis, A. and Brownell, A. G.**, Collagen Biosynthesis, *CRC Critical Reviews in Biochemistry*, 2(4), 417, 1975.

134. **Verdan, C., Ed.,** *Tendon Surgery of the Hand* Churchill Livingstone, Edinburgh, 1979, 1.
135. **Viering,** in *On the Reparative Processes in Human Tendons After Subcutaneous Division for the Cure of Deformities*, Adams, W., J. Churchill, London, 1860.
136. **Viidik, A.,** Thermal contraction-relaxation and dissolution of rat tail tendon at different ages. *Aktuelle Gerontologie*, 7, 493, 1977.
137. **Viidik, A., Andreassen, T., Busted, N., and Oxlund, H.,** Maturation and ageing of collagen — some functional aspects. *Geron: Helsinki*, 21, 16, 1977.
138. **Viidik, A.,** Connective tissues — possible implications of the temporal changes for the ageing processes. *Mech. Ageing Dev.*, 9, 267, 1979.
139. **Vitto, J. and Lichtenstein, J. R.,** Removal of amino-terminal and carboxy-terminal extension peptides from procollagen during synthesis of chick embryo tendon collagen. *Biochem. Biophys. Res. Commun.*, 71(1), 60, 1976.
140. **Vogel, H. G.,** Influence of maturation and age on mechanical and biochemical parameters of various organs in the rat. *Connect. Tissue Res.*, 6, 161, 1978.
141. **Walker, P., Amstutz, H. C., and Rubinfield, M.,** Canine tendon studies. II. Biomechanical evaluation of normal and regrown canine tendons. *J. Biomed. Mat. Res.*, 10, 61, 1976.
142. **Warwick, R. and Williams, P. L., Eds.,** *Gray's Anatomy*, 35th ed., Longman Group, Saunders, Philadelphia, 1973, Ch. 4, 454.
143. **Winckler, G.,** Anatomy of tendons and sheaths, in *Tendon Surgery of the Hand* Verdan, C., Ed. Churchill Livingstone: Edinburgh, 1979, 8.
144. **Wynne-Davies, R.,** Acetabular dysplasia and familial joint laxity; two etiogical factors in congenital dislocation of the hip. *J. Bone J. Surg.*, 52B, 704, 1970.
145. **Zernicke, R. F., Garhammer, J., and Jobe, F. W.,** Human patellar tendon rupture. *J. Bone J. Surg.*, 59A, 179, 1977.
146. **Zielinkski, J. R., Rudnicka, A., and Rymdzconek, M.,** Tensile strengths and Young's modulus of the posterior cruciate ligament in the dog. *Folia Morphologia* (Warsaw) 24, 249, 1975.

Chapter 9

STRUCTURAL AND MECHANISTIC CONSIDERATIONS IN THE STRAIN-RELATED ELECTRICAL BEHAVIOR OF BONE

G. W. Hastings

TABLE OF CONTENTS

I. INTRODUCTION

In a previous publication[1] the author and colleagues described results of an investigation which demonstrated that bone has the character of a ferro-electric material and therefore is by definition piezo-electric with respect to the transduction of mechanical forces. The interesting consequence of this is that the charged dipoles, which give rise to the electric charge produced when bone is strained, must be arranged in regions or domains. The orientation within each domain may not be uniform but can be changed when an electric field is applied. Further support is given to this view by the observation [2] that bone electrets may be formed by various methods, and this again implies an ordered domain structure.

The actual origin of strain-related potentials is not, however, determined by establishing that there are charge-producting structures present, although the concept does illustrate how mechanical deformation of such an electrically ordered region will produce a nonequilibrium effect and hence net potential. Nor does the observation of this phenomenon ascribe any physiological function for the potential that is produced. The potentials observed may be solely descriptive of a mechanical distortion that has occurred, in that sense monitoring the distortion, but being therefore of secondary importance. The effect may, however, serve to limit excessive deformation by the action of charge repulsion at macromolecular or cellular level. Alternatively, the potential may direct, control, or initiate cellular activity, as it seems that an applied electrical field is able to do. This then imparts to it a primary role in bone growth or repair processes.

In this chapter, certain observations by ourselves and other groups will be discussed and an attempt made to draw certain tentative conclusions regarding structural features and their relationship to bioelectric phenomena and to growth or healing in bone.

II. THE LOCATION OF DOMAINS

The previous paper[1] reported the finding that bone exhibits ferro-electric characteristics and therefore that it is a piezo-electric material. However, the classical piezo-electric theory does not apply to bone, and it may not be correct to equate domains with structural units. There is a recognized structural component in bone, the osteon, representing distinctive regions of order quite distinct from adjacent ones in which collagen arrangement is different, Pollack et al.[3] have studied the change in potential by microelectrodes as the point of measurement is moved across the surface of a stressed bone from the tension to the compression side. The significant observation was that whereas the potential gradient between the ends of the complete sample may be 0.01 V $cm^{-}1$, in the vicinity of the Haversian canal the field gradient may be forty times the average value. In tension the field direction is away from the Haversian canal, and in compression it is directed towards the Haversian canal; i.e. the field direction is reversed. This is despite the fact that the stress gradient across the whole sample is constant. Near the neutral axis the voltage gradient is zero. The large fields measured in the region of the Haversian canals are comparable in magnitude with those used in direct current induced osteogenesis in vivo. A further interesting observation is that under compression the direction of the field toward the Haversian canal is equivalent to making the inside of the canal correspond to a negatively charged electrode. Bone will grow in response to a negative potential of the magnitude measured here. It is known that bone will grow in response to compression and, as Frost [4] observes, grows towards the concavity of a long bone so that the concavity is eliminated under normal conditions. Pauwels[5] has utilized this principle of growth in response to physiologically acceptable levels of compression in his work on osteotomy of the proximal femur as a treatment for osteoarthrosis, and it is of course the utilization of the Wolff's Law principle.

The results of Pollack et al. show that the net surface-surface potential is a summation of the potentials at the individual osteons and that across a sample the resultant magnitude will be the same irrespective of loading direction. The sign only will change on a surface as the loading changes from tension to compression and hence as the sample is end-to-end rotated. Rotation about the axis of the Haversian canal shows that the field is directed radially. Change in thickness of a sample will change the size of the potential unlike a classical piezo-electric material. An essential difference between microscopic and macroscopic behavior is shown by these experiments. The macroscopic potential is dependent on stress gradient and will be zero for zero stress gradient. This is not the case at the microscopic osteonal level, where potentials will still be produced. The work also illustrates the way in which at the local level the osteon amplifies the effect of macroscopic strain.

A. The Osteonic Structure of Bone

It is apparent that the osteon is a fundamental unit of organization in bone structure, and it is therefore appropriate to consider whether it is also a fundamental unit in electrical behavior. The difference between macroscopic behavior and microscopic osteonic electrical characteristics has been discussed above.

Three types of osteon — dark, bright, and intermediate — have been described and their structures examined. It has generally been proposed that in the dark osteon the collagen fibers are mainly found to be arranged longitudinally whereas bright osteons have circumferential fibers in addition. In the intermediate osteons longitudinal and steeply oblique fibers predominate, with fibers of other orientations also present. Wide-angle x-ray diffraction examination of dark osteons indicates the c-axis of the inorganic crystals to be along the osteon axis more so than for the bright osteons. If it be accepted that the apatite crystal is aligned with its c-axis along the collagen fibril direction, then this reinforces the SEM evidence. Katz[6] records an observation that in one study no osteons were found in which fiber orientation in adjacent lamellae alternated from longitudinal to circumferential arrangements. However, the above results show that the structure is by no means uniform, and this has been emphasized by Boyde,[7] who cautions against drawing mechanical conclusions from a superficial consideration of bone fiber arrangements. The integrity of the inorganic framework of deproteinized bone shows the continuity of bone crystal, and the epitaxy between microfibrils and crystals is certainly more complex than the original views suggested.

ElMessiery[8] has made a mathematical analysis of the piezo-electric characteristics of a single osteon, making the assumption that the collagen orientations in successive lamellae are differentiated by a wrapping angle (α) between them, which he proposed was either $\pm 90°$ or $\pm 180°$. He considered an infinitesimal cubic element of collagen fiber in one lamella, assigned axes, and treated it as a piezo-electric crystal belonging to the D^6 hexagonal crystal group,[9,10] in which case there are only two non-zero moduli, d_{14} and d_{25}.

The consideration that the cross-section of the osteon is oval rather than circular$_{10}$ led ElMessiery to modify his analysis. $d_{14}\ \alpha$ and $-d_{25}\alpha$ are not now identical, and d_{36} results must be considered from another component of polarization. The result was to show that the SGP tensor is now not only dependent on the wrapping angle but on the ratio of two axes of the elliptical cross-section. If cylindrical coordinates are used instead of cartesian, the three polarization components of an osteon may be expressed as:

$$Pr = d_{25}\ (\alpha)\ \text{osteon}\ T_{\beta z} = d_1 T_{\beta z}$$

$$P_\beta = d_{25}\ (\alpha)\ \text{osteon}\ T_{rz} = d_2 T_{rz}$$

$$P_z = d_{36}\ (\alpha)\ \text{osteon}\ T_{r\beta} = d_3 T_{r\beta}$$

with $d_{1,2,3}$ being the non-zero piezo-electric coefficients and $T_{\beta z, rz, r\beta}$ the cylindrical representation of the shear stress components.

This work was a useful attempt to describe the osteon as the basic piezo-electric unit in bone and to relate its behavior to that of the collagen fibers of which it is comprised, as well as to the angle of these in one lamella compared with another and to the cross-section of the osteon. He was able to conclude that the simple osteon can be described as a single piezoelectric crystal with piezo-electric coefficients well related to those of its component collagen fibers. Unfortunately, the treatment of wrapping angle is inevitably simplistic and even within one lamella there is variation. Although not contributing to the electrical effect directly, the inorganic crystals may also have some effect on collagen-collagen interactions.

III. ELECTROCHEMICAL CONSIDERATIONS

A considerable study of electrochemical effects in cell function has been made by Pilla,[11] using pulsed magnetic fields. One of the hypotheses has been that changes in electrochemical environment around a cell will lead to modifications in the structure of its charged surface regions. This could be brought about by changing the concentration of bound ions or dipoles, which may be accompanied by conformational changes of molecular species in the membrane structure. Hence, Pilla states that cell regulation involves both potential and kinetic factors associated with specific biochemical processes. Local structural modifications can take place in the cell membrane, and these may be triggered by very specific pulses, which thus gives the possibility of selectively altering the kinetics of function by appropriate choice of current waveforms and repetition rate. He has calculated that the change in charge required is only microcoulomb per cm^2 — i.e. microamp per cm^2 current intensity — and to ensure that the pulses have spatial and time uniformity at cell level, electromagnetic induction of current was considered the best method. He was able to demonstrate "de-differentiation" effects in frog red blood cells but, interestingly, in a normal ionic environment, morphology changes corresponding to functional change "de-differentiation" were not observed, and pulsed induced current did not effect changes. It appears that the induced current modulates the rate of response to the cells' existing environment, i.e., unless the cells are in an abnormal environment or are capable of functional response due to the changing microenvironment (e.g. in trauma), then pulsed current is not able to modulate a condition that is not extant. Variations in chromatin-DNA structure were observed, and an increase or decrease in rate was dependent upon signal frequency and waveform. Certain waveforms may affect transport kinetics of metal ions into the intracellular fluid, and the increased concentration of these ions will alter the DNA-metal ion interactions in the cell, which appear to be of significance in cell differentiation and pathology. Cell surface changes and ionic availability are thus the two important processes which seem to be capable of variation by means of induced currents. Pilla's concept is thus that change in the electrochemical microenvironment of the cell can cause modification of its charged surface regions, possibly by changing the concentration of specifically bound ions or dipoles. This may be accompanied by conformational modification in the molecular entities in the membrane structure. He concludes that the regulatory interactions at a cell surface may be dependent both on electric potential and chemical kinetics associated with the specific biochemical events required for the various processes.

Molecular movements of cell membranes components are undoubtedly influenced by environment, and it seems reasonable to expect that local variations in the concentrations of charged moieties can be effected by electrochemical processes.

It should be noted that the studies referred to above are non-Faradaic in nature, in contrast to those of Black and Brighton.[12] The effect of implanted electrodes through which 5 to 20 microamperes direct current is passed directly into the 'fracture site', the electrode being the cathode, is to promote local pO_2 reduction and pH increase. Stimulation of healing is produced by a physical chemical route as an intermediate stage. There is thus an electrode reaction to be considered also in this case.

Although the effects produced by Pilla and by Brighton differ in that the former have a direct electrochemical initiation whereas the latter involve the electrode reactions and the possibility of an intermediate physico-chemical event, the resultant perturbation of cell processes or membrane structure may be ultimately the same.

IV. INORGANIC BIOCHEMISTRY

It is now necessary to consider inorganic chemical factors, and these have been discussed comprehensively by Williams.[13] He starts from the premise that the viable cell is in a steady state with respect to its environment and all the dynamics — chemical, mechanical, and electrical — are under control. Energy distribution is critical and control of this utilizes both small molecule diffusion and transfer in a polymer structure. In fact, flows of small molecules are of primary importance in the elucidation of control mechanisms. Controls will be related both to steady state conditions and to the need for rapid response to external change or to a message, and these latter are referred to as switch controls by Williams. Different processes will require different pathways and, rather than use phase considerations, he considers the importance of micro-domains which had been previously postulated as the basis for organization. Each domain can be separately energized at local levels within an otherwise organized layer, hence permitting local circuits. The parallel thought with Pilla is of note and one can begin to see ways in which an electric field, external to the cell, could influence its internal control processes, via small ion flux changes or membrane structure variations, and also its interaction with the environment or to the initial reaction. Inter-relationship between the control systems will provide a feedback system.

There are various possibilities for control of the cell's steady state. Transfer of the major metabolic elements by the action of co-enzymes gives a very selective control, regulated by redox potentials. Energy flow in the cell needed for synthetic processes is usually controlled by phosphate systems, particularly ATP. The above systems are essentially for relatively unrestricted phases, but within cell particles, movement of elements is by entrapped co-enzymes utilizing covalent bond formation. Mechanisms of transport through the cell membrane are a further source of control steps, and additionally there is control of mechanical (i.e. shape) and electrical steady states. Electrical or mechanical effects on the membrane can lead to conformational changes in the assemblies of proteins. Osmotic and local control is a further factor, but Williams has concluded that local circuits for control are the more important.

When internal structures, vesicles, are introduced into a cell, further membranes are introduced, and the possibilities of control are increased.

As well as the slower steady state controls there are rapid responses to stimuli, and different switches must be invoked. Chemical, mechanical, and electrochemical switches are possible. The former is based on rapid, pulsed release of chemicals, e.g. the production of c-AMP from ATP activated by an enzyme produced by change in shape of a macromolecule from an inactive to an active form. Mechanical switches offer a range of controls, e.g. muscle action and development of tension. The latter switch involves charge carriage across membranes using $K+$ and $Na+$ ions. It is possible for a change in sodium ion diffusion rate, an $Na+$ current, to open a channel for calcium ion influx, which has a large chemical effect leading to release of stored agents. Protons can act in a similar manner and very localized pH changes exert profound effects.

Hence, it is seen, as Williams concludes, that change in rate of flow of chemicals is the significant factor in transmission of signals, and these changes can lead to changes in cell structure.

This brief extraction of some salient points from the first article by Williams, which concentrated on inorganic chemical entities, has indicated several areas which could be affected by input of electrical energy or by electrode reactions. Since movement or rate of flow are suggested to be of first rank importance, these could easily be altered by changes in electrical environment. Change in equilibrium charge potential across or around an osteon unit, resulting from mechanical strain, will lead to a change in electrical conditions and this could rapidly alter cell structure and function. Cell membranes comprise protein molecules,

Table 1
SWITCHES CONTROLLING CELLULAR ACTION

Chemical external internal	ATP, Ca^{++} flux
Mechanical stress Mechanical strain	Mediated by chemical change resulting from external forces
Electrochemical ion diffusion	Electrode stimulated
Redox reactions	Electrode stimulated
Proton migration	Proton tunneling

and conformational changes influenced by various factors are possible in these. All macromolecule chains have considerable mobility influenced by free energy considerations. Theories of rubber elasticity and visoelasticity in macromolecules permit some understanding of the way in which chain mobility is permitted, and the manner by which different chain segments influence molecular rigidity. External influences also need consideration, and this includes perturbation of structure due to electrical potentials.

As Williams points out in his second paper, the observation that there are movements in enzyme molecular structure on binding of their substrates indicates a possible control feedback mechanism. If a relatively minor structural change only is needed at an active site in the presence or absence of a control chemical binding at a second site, then the protein really needs only two conformations for switching control to be achieved. Furthermore, strain factors may be important, for if relief of strain in a chain is brought about by chemical reaction, this will enhance the progress of that reaction by resulting in lower energy. In terms of bone remodeling, this latter factor may indicate the possibility of calcification processes being triggered by mechanical effects, i.e. strain, directly — without the necessity for electricity as an intermediate. It should, of course, be remembered that all biological systems are characterized by movement, of energy or of material, and represent a metastable situation, and that the protein 'structures' of a membrane or of an osteon, are continually variable as a whole and at local levels. Hence, there are several pathways across a membrane, energized locally, but the energization of one alters all the structures. Although proteins exert control, their function is also controlled by small molecules and ions. He draws attention to calcium flux, which being a divalentcation, is a particularly effective control action in structuring and restricting chain conformations, and any process which involves calcium flux may be highly significant. The function of electrical energy may be to vary this flux.

V. DISCUSSION

Although electrical effects are a clearly demonstrable property of living and of dry bone, it is not by any means certain that those potentials produced as a consequence of mechanical strain exert any biological effect. The fact that externally applied electricity appears to have an effect on the healing of fracture non-union may be quite irrelevant to the study of strain-generated potentials.

Several of those studying the use of applied fields refer to a "window" or "trigger" effect, indicating that a very specific form of electrical signal is required, and this appears to be confirmed by the work of Fitton-Jackson.[14] She recorded differences in the growth of chick embryo explants grown in tissue culture when the nature of the pulse was changed. Pilla drew attention to the ionic environment rather than the current itself as being important, and it is the changed ionic environment acting on an abnormal situation that causes growth. Korostoff,[10] in studies on stress-generated potentials, also refers to a trigger effective on the

Table 2
EFFECTS OF STRAIN GENERATED POTENTIALS

Cells	Ionic environment
	Ion flux
	Energy of activation reduction
	Strain relief effects on membranes
Macromolecules	Alignment/orientation on electrical field gradients
	Control over strain
	Control over transport phenomena
	Change of reactivity by orientation

internal polarization of bone and proposes that this is based on the net molecular surface charge of bone. This, as has been pointed out above, depends on the summation of effects across individual osteons. Hence, in unstimulated bone, i.e. bone in vivo strained within normal physiological limits, the physical form of the strain-generated potential affecting tissue growth could also be expected to be critical. Osteon shape, size, or alignment may produce this specific form, and therefore the location of the osteon on the stress gradient profile will determine the nature of the resultant field and thus have direct feedback relationship to growth or resorption of bone.

If in vivo strain-generated potentials have a control function over growth levels, two levels of control may be considered, viz., at cellular level and at the macromolecular level. There are also positional or orientation effects. (See Table 2).

At cellular level, change in ionic environment or ionic flux possibly associated also with membrane structure changes may be expected to amend cell function. The cells may act in osteoblastic or osteoclastic activity, although it is hard to see how current or its direction could affect this unless the direction of ion flux was reversible by this means or that different ions became involved at different levels of electrical activation. Differing energies of activation for cell processes were referred to by Williams, and hence it is possible that certain control pathways are activated by the magnitude or duration of the field generated in vivo.

At the macromolecular level there are several possibilities. If mechanical deformation itself can also affect cell membrane behavior directly, in addition to its role in potential formation, then the actual level and direction of strain is important and it may act independently or synergistically with electrical effects. The action of strain relief in chemical reaction has been referred to with its effect on membrane structure and function. The alignment of collagen fibrils as well as their rate of production could be influenced by electrical fields, and levels of fields in the osteon are sufficiently high to be a consideration. Mechanical properties of bone are dependent upon its orientation with respect to loading directions, and hence the anisotropic nature of the mechanical deformation produced by an applied load produces mechanically and positionally determined controls upon strain-related phenomena, whether of cell reaction or electric potential. The physical location of an osteon in the strained bone will determine the magnitude and direction of its deformation and hence of its electrical response. This will be modified as the bone grows or changes its alignment and hence this acts as a control loop. Since bone shows viscoelastic behavior (i.e. time dependence of mechanical properties), the delayed relief of strain may extend the period over which electrical potentials can be generated and also over which molecular distortion occurs. Hence, the possibility of ion binding is extended and thus of calcification. Energy storage properties, viscoelastic or ferro-electric, therefore acquire a physiological significance. (See Table 3).

There is a clinical relevance in these basic studies. It is increasingly being realized that fractures heal by different routes when the fracture is treated conservatively in plaster or fixed internally by rigid means. In the former case external bridging callus is produced to

Table 3
MECHANO-ELECTRICAL CONTROL

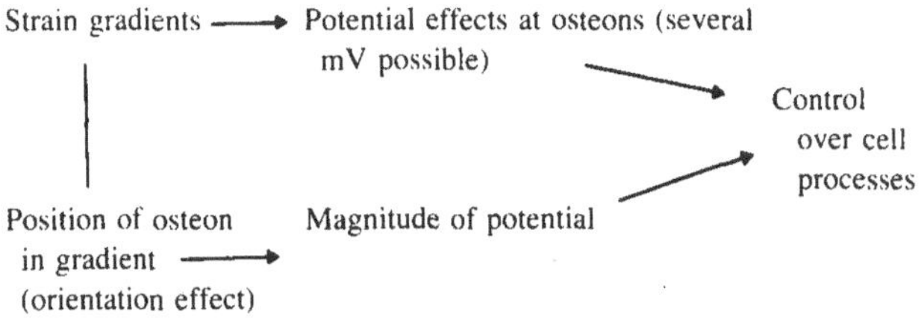

Table 4
FACTORS IN FRACTURE FIXATION

	Increase	Decrease
Stability of fixation	Union of fracture	Delayed or non-union
	Strain reduction	Strain increase
Rigidity of fixation	Less strain in bone.	More strain in bone
	Cortex-cortex healing (primary union)	Callus healing
Strain tolerance level	Effect to be investigated	
	Balance between primary union, callus, non-union; electrical effects dependent on orientation	

give early consolidation. In adequate rigid internal fixation, callus is not seen and healing occurs by a late process or cortex-cortex growth referred to as primary bone union. (See Table 4.) The advantages are being realized in certain cases of less rigid internal fixation which nevertheless still provides for stability of the fracture. In this case external callus is seen.[15] It appears that less rigid fixation permits strain at the fracture site and this "triggers" callus formation. Callus production is an early phenomenon and appears to be stimulated by weight bearing. The out-of-balance forces pertaining when a fracture is present are effective, but the practically allowable limits of these are not known. (See Figure 1).

If the basic studies on strain-generated potentials can help to define more precisely the tolerance of living bone for strain in normal conditions and when a fracture is present, this may prove in the long term to be a more valuable service in terms of patient treatment than electrical stimulation, which is a salvage procedure. Extension to pathological conditions in bone is then another possibility.

VI. CONCLUSIONS

The electrical domain structure of bone having been proposed as a result of its ferro-electric nature, the physical location and significance of these mechano-electrically active regions remains to be demonstrated. The microscopic response is shown to be different from macroscopic observations, and the osteon is seen to be a key structure in this. If, as seems to be demonstrated by induced field and electrochemical effects, there is a response 'window' or 'trigger' then in normal bone in vivo the position of the osteon and its alignment on the applied stress gradient may provide the trigger potential relevant to the Wolff's Law response required. In this way mechanical effects are mediated by electrical response and this is structurally determined.

Mechanical effects may act directly by causing membrane deformation with consequent change in its function.

A further mechanical factor relates to the antisotropic nature of bone and to its viscoelastic

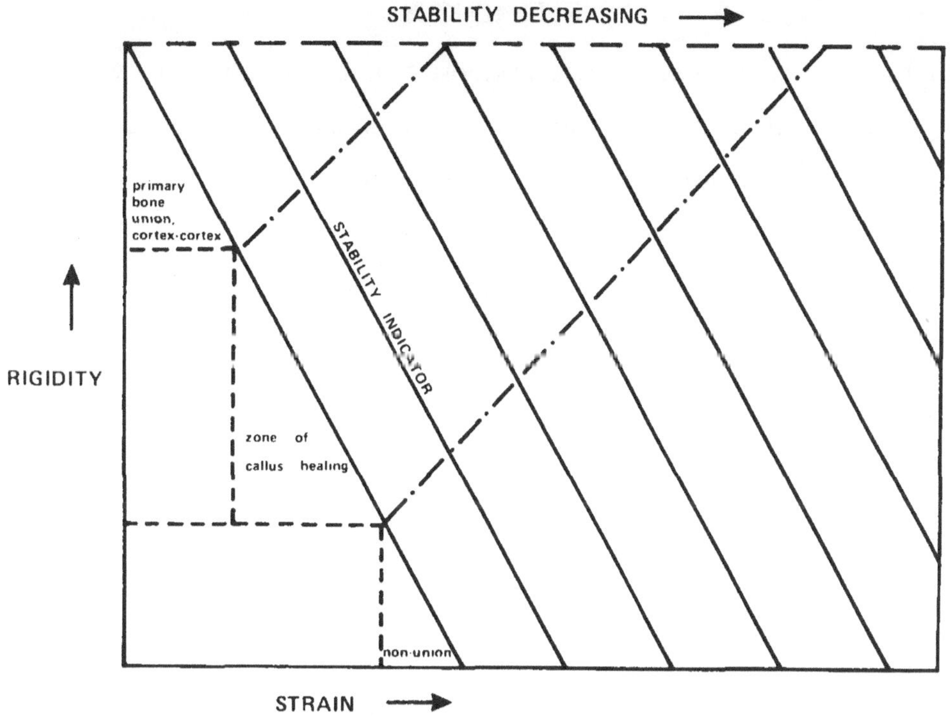

FIGURE 1. Stability of fracture fixation related to rigidity and mechanical strain. (Presented by G. W. Hastings at Conference on Polymers in Medicine, Porto Cervo, Sardinia, May, 1982).

behavior. This emphasizes the orientational nature of control processes and indicates that local intra-osteon variations are possible. Bone will store energy, electrical as instanced by the ferro-electric hysteresis, and mechanical in its time-dependent response. This may be seen as part of the defense against response to sudden changes so that they are "smoothed out"; i.e., as Frost says, only time-averaged forces are important, or it may be the mechanism by which the duration available for change to occur is extended.

It must be remembered, as Williams emphasizes, that all structures interact with their environment and in so doing afford several reaction pathways and possibilities for control function. Ion flux is important in this and can be varied by electric phenomena.

Finally, it is to be reiterated that the long-term value of these studies may be in determining the parameters for treatment of trauma rather than in the use of stimulation in salvage.

ACKNOWLEDGMENTS

The author wishes to thank various colleagues for helpful discussions and in particular the main authors referred to, whose writings have 'stimulated' this discussion.

REFERENCES

1. **Hastings, G. W., ElMessiery, M. A., and Rakowski, S.,** Mechano-electrical properties of bone. *Biomaterials,* 2, 225, 1981.
2. **Andrabi, W. H. and Behari, J.,** Formation of bone electrets and their charge decay characteristics. *Biomaterials,* 2, 120, 1981.
3. **Pollack, S. R., Korostoff, E., Starkebaum, W., and Iannicone, W.,** Microelectrode studies of stress generated potentials in bone, in *Electrical Properties of Bone and Cartilage.* Brighton, C. T., Black, J., and Pollack, S. R., Eds. Grune & Stratton, New York, 1979, 69.
4. **Frost, H. M.,** *Orthopaedic Biomechanics.* Springfield: Charles C Thomas, Springfield, Ill., 1973.
5. **Pauwels, F.,** *Biomechanics of the normal and diseased hip.* Springer-Verlag, Berlin, 1976.
6. **Katz, J. L.,** The structure and biomechanics of bone in Vincent, J. F. V., and Currey, J. O., Eds., *The Mechanical Properties of Biological Materials,* Cambridge University Press, Cambridge, 1980, 137.
7. **Boyde, A.,**The texture of bone. Biological Engineering Society Biomaterials Group meeting "Biomechanics of Bone". Queen Mary College, London, 1981.
8. **ElMessiery, M. A. I.,** The origin of electricity in bone with reference to its electromechanical behaviour. Ph.D. thesis, North Staffordshire Polytechnic, Stoke on Trent, U.K., 1979.
9. **Guzelsu, N.,** A piezoelectric model for dry bone tissue. *J. Biomech.* 11, 259, 1978.
10. **Korostoff, E.,** Model for characterising stress generated potentials, in *Electrical Properties of Bone and Cartilage.* Brighton, C. T., Black, J., and Pollack, S. R., Eds. Grune & Stratton. New York, 1979, 95.
11. **Pilla, A. A.,** Electrochemical information transfer and its possible role in the control of cell function, *Electrical Properties of Bone and Cartilage.* Brighton, C. T., Black, J., and Pollack, S. R., Eds., Grune & Stratton, New York, 1980, 455.
12. **Black, J. and Brighton, C. T.,** Mechanisms of stimulation and osteogenesis by direct current, in *Electrical Properties of Bone and Cartilage,* Black, J., Brighton, C. T., and Pollack, S. R., Eds., Grune & Stratton, New York, 1980, 215.
13. **Williams, R. J. P.,** On first looking into nature's chemistry. Part I. The role of small molecules and ions: the transport of the elements. *Chem. Soc. Rev.* 9, 281, 1980. Also part II. The role of large molecules, especially proteins. *Chem. Soc. Rev.* 9, 325, 1980.
14. **Fitton-Jackson, S., Farndale, R. F., and Jones, D. B.,** The response of skeletal tissues grown in organ culture to pulsed magnetic fields. Biological Engineering Society Symposium, Stoke on Trent, May, 1980.
15. **Tayton, K., Johnson-Nurse, C., McKibbin, B., Bradley, J., and Hastings, G. W.,** Results of preliminary trials using semi-rigid CFRP plates for human fracture fixation. *J. Bone J. Surg, Vol.,* 64B, 105, 1982.

INDEX

A

B

C

D

E

F

M

N

O

P

R

S

T

U

V

W

X

Y